A Background to Primary School Science

A Background to Primary School Science

Neville Fletcher

Research School of Physics and Engineering

The Australian National University

Published by ANU eView
The Australian National University
Canberra ACT 0200, Australia
Email: anuepress@anu.edu.au
This title is also available online at: http://eview.anu.edu.au

---

National Library of Australia Cataloguing-in-Publication entry

Author: Fletcher, N. H. (Neville Horner)

Title: A background to primary school science / Neville Fletcher.

ISBN: 9781921934049 (pbk.) 9781921934056 (ebook)

Subjects: Science--Study and teaching (Primary)--Handbooks, manuals, etc.

Dewey Number: 371.33

All rights reserved. No part of this publication may be reproduced, stored in a retrieval system or transmitted in any form or by any means, electronic, mechanical, photocopying or otherwise, without the prior permission of the publisher.

Typeset by the author in LaTeX
Drawings by the author

Cover design by ANU E Press

This edition © 2011 ANU eView

Preface to Revised Edition

During the year 2001 the original edition of this book was used by Dr Joan Robson and her Teacher Education students at the Signadou Campus of the Australian Catholic University in Canberra. To them all, I offer my thanks. Encouraged by the results of that trial, I have made several minor corrections to the text and have also added some new material in appendices that will, I hope, make the book useful to an even wider range of students and teachers.

As in the first edition, I have retained hand-drawn figures throughout the book, though it would have been easy to replace these by nice clean computer-drawn diagrams and pictures. The reason for this is that the originals have proved popular by indicating the sort of thing that teachers can easily draw in class.

I hope that the book will prove useful, not only to students undertaking preparation for teaching, but also to those who have spent many years in the profession but are now faced with teaching scientific material for which they were never adequately prepared. To all of you, my best wishes in an important task.

January 2002   Neville Fletcher

Preface to 1995 Edition

The Australian Academy of Science has had a long standing interest in the provision of science education to Australian school students. This concern began nearly thirty years ago with the introduction of the text *Biological Science: The Web of Life* which revolutionised biology teaching in the senior years of Australian schools, and has since extended to cover nearly all areas of science at senior high school level. There has been a new text in Biology, and major texts in Chemistry, Earth Sciences, Mathematics, Public Health and Environmental Science. Most of these programs are supplemented by the provision of periodical newsletters and by visits from field officers.

Recognising that skills and attitudes in science are acquired at an early age, the Academy is launching in 1994 a major program in science education at primary school level under the title *Primary Investigations*. This development comes at a time when there is a national focus on the goals of education at all levels, and incorporates the best features of the National Statements and Profiles in science, as well as providing scientific background necessary for studies in technology and in environmental science. These materials take account of the fact that many primary school teachers have been inadequately prepared, in their professional education, to teach science with confidence. The course materials therefore provide all the necessary background information as well as detailed instructions for conducting the hands-on experimentation that is a key feature of the course. Lower secondary school, however, remains an area to which the Academy has yet to pay detailed attention, but which will be the subject of detailed consideration in coming years.

Since science teachers at primary and lower secondary levels come from a wide range of backgrounds and many of them have not studied all the science subjects in detail, I have tried here to provide a broad survey of what is known about the world from a scientific perspective. Clearly I have had to be selective in what is covered, and in this I have been influenced by the present and likely future structure and content of science courses at primary and lower secondary levels in Australian schools. The material presented, however, is not what teachers should expect to teach, but rather a much broader background that should help them to place school science and technology studies into a global context.

I hope that this little book will prove useful to teachers using the Academy course materials, and also to those teaching in other ways. Perhaps it may also find a more general readership among parents and secondary-school students. Science is an important part of our heritage and a vital practical component of our collective future. Our children have a need to understand its basis and to appreciate both its relevance and its intellectual excitement.

March 1994   Neville Fletcher

Contents

1. This Thing Called Science   **1**
    The nature of science 1
    Analytic and holistic science 4
    Women and science 5
    The applications of science 5

2. Getting the Facts   **7**
    Experiments in physical science 8
    Experiments in biological science 10
    Where experiment is impossible 11

3. Change and Energy   **13**
    Energy and heat 13
    Stored energy 16
    Light energy and sound energy 21
    Wasted energy 23

4. Hearing and Seeing   **25**
    Sound and hearing 25
    Light and seeing 28
    Lenses and mirrors 32

5. The Universe and Everything   **36**
    The solar system 38
    A modern view of the universe 41
    The future of us all 43

6. This Earth of Ours — 44
   The structure of the Earth 45
   Landforms 47
   Rocks and fossils 49
   Minerals 50
   Fossil fuels 51
   The oceans and ice caps 52

7. Weather and Climate — 55
   Driving the weather 56
   Clouds and rain 60
   Climate 62
   Climate change 63
   Atmospheric pollution 64

8. Magnetism and Electricity — 66
   Magnetism 67
   Electricity 69

9. Atoms, Molecules and Chemistry — 74
   Atoms and elements 75
   Molecules 76
   Crystals 77
   Beyond atoms 80
   Chemistry and cooking 81
   Soap, detergents and dirt 83
   Natural and synthetic chemicals 84

10. Useful Materials — 87
   Natural materials 87
   Ceramics 88
   Metals 89
   Plastics 90
   Foams and fabrics 91
   Materials and design 91

11. Life and Living — 93
   The origin of life 94
   The variety of life 98

12. The World of Plants — 104
- Photosynthesis — 105
- How plants live — 106
- Reproduction — 107
- Plants as energy — 109

13. The World of Animals — 111
- The history of animals — 111
- The animal body — 113
- The energy of humans — 115
- Nerves and the brain — 116
- Consciousness — 117

14. Living with Nature — 119
- Shelter — 119
- Food — 120
- Managed ecosystems — 122
- The problems of overpopulation — 123
- Water — 125
- Energy — 126

15. Materials and Structures — 130
- Compressive strength and design — 130
- Tensile strength and design — 132
- Structures and stability — 136
- Biological structures — 137

16. Tools and Machines — 140
- Effectiveness — 141
- Efficiency — 141
- Machines — 143
- Engines — 144
- Controlling machines — 145

17. Communicating Information — 147
- Information — 148
- Communications — 149
- Computers — 151

Appendix A: Scientific method   **155**
    The scientific method 155
    Hypotheses and postulates 156
    Definitions and laws 157
    Uncertainties and statistics 158
    Revolutionary ideas 160
    The community of science 162

Appendix B: Scientific units   **163**
    SI units 163
    Special cases 165

Appendix C: Science and Ethics   **167**
    Nuclear power 168
    Genetic engineering 169
    Cloning 170

Bibliography   **172**

Index   **176**

1

This Thing Called Science

"The whole of science is nothing more than a refinement of everyday thinking." Albert Einstein

To many people science is a rather mysterious activity carried out by white-coated researchers in their laboratories, using strange and expensive machines, and has little to do with everyday life. It is certainly true that some of science is like that, and is rather hard for ordinary people to understand, but a great deal of science deals with things are that around us all the time, and some understanding of scientific methods and principles will help us to make sense of many important questions. Later chapters of this little book will tell the reader something about the important principles and applications of science, but first let us look at what science is and how scientists go about their work. I will also say a little about the applications of science, for these are of immense importance to our present lifestyle and to the future of our children.

The Nature of Science

While it might be nice to think that science is searching for "the truth" about the universe and the way it works, the idea of truth is a very elusive one and science really aims to do something quite different. What scientists try to do is to construct models—usually mental models, which are called theories—that mimic the way in which particular bits of nature operate and thus allow one to predict what is going to happen. A theory is useful only if it gives predictions that are reliable—science aims to be "reliable knowledge".

It is almost too much to hope that any theoretical model could predict the behaviour of any real system with absolute accuracy in all its details, but there is a clear distinction between good models and bad models, and this distinction can be made quite easily by putting the models to experimental test—by examining their predictions against what actually happens. It is a vitally important part of science that its theories are testable. Any set of beliefs that does not lead to testable predictions is simply dogma—it may be comforting but it is of no practical value.

As science develops it quickly discards really bad models that give incorrect predictions in simple cases, but much more interesting is what happens to really good models that work for most cases but fail under extreme circumstances. Such models are too good to discard, and what usually happens is that their limitations are realised and they are used only within these limits, new and generally more complex models being required in unusual situations. A good example is the set of principles and general theory developed by the English mathematician Isaac Newton (1643–1727) some three hundred years ago to describe the motions of the planets, and incidentally the motions of all other objects. Newton's mathematical theory, based upon earlier observations by Galileo Galilei (1564–1642), describe almost exactly all these phenomena and allow us to predict the orbits of comets, to design jet aircraft, and to navigate a space probe through the solar system to the outer planets. No-one in their right minds would now think of using any other theory to make these calculations. But it is known that Newton's theory gives incorrect answers when applied to things moving at speeds approaching that of light—an immense 300,000 kilometres per second. Even at one tenth the speed of light the error in a Newtonian prediction is about 1 percent, but at ordinary speeds the necessary correction is completely negligible. Einstein's new theory of relativity, developed about eighty years ago, gives results that are exactly correct in these cases, as far as we can measure them, but of course its predictions must agree with Newton's theory for ordinary speeds. Newton's theory also breaks down when we come to calculate the behaviour of atoms, but again that is a new opportunity for advance—the development of quantum mechanics—not a destructions of existing understanding.

Journalists and sociologists of science are addicted to the rhetoric of revolution and speak of relativity and quantum mechanics as having "overthrown" Newton's theories, but that is simply incorrect—scientists use whichever theory or model is most appropriate for the task in hand, as long as it gives reliable results.

It is important to look at the way in which scientists go about building their models and theories, because there is nothing strange about it and it can provide a model for the way in which we ought to proceed in much more everyday tasks. The primary principles are simply honesty and lack of secrecy, coupled to a sceptical view of matters so that nothing is really believed until it can be clearly and repeatably demonstrated.

Basically what happens in any field of science is something like the following. By watching what happens in nature we observe certain occurrences for which there is no obvious explanation. Sometimes an explanation may occur to an individual as a flash of insight, but often it is necessary to carry out experiments to remove complicating influences and show the interesting events in their simplest possible form, after which it may be easier to guess an explanation. Whether we have guessed one or many possible explanations, we must now put the matter to the test by making predictions based on those explanations, discarding those that give the wrong answers when tested. The whole thing is then discussed with other scientists by publishing a paper about it, describing the experiments and their results and the conclusions drawn from them. Because most things these days are rather complicated, other scientists will try to repeat the experiments themselves to make sure that the observations were sound, and will try to think of other simpler explanations of what is going on. When everyone is agreed, then the whole thing becomes an accepted part of science—not fixed for ever, but the most reliable knowledge available at the time.

All this sounds straightforward, but there may be many complications. The performing of experiments often requires a great deal of skill and special equipment that has itself been developed over long periods of time by scientists studying similar problems. The working out of results and the construction of theories may require complicated mathematics and long hours of computing, and scientists must be familiar with all these techniques. But one can't just turn a handle and manufacture scientific understanding—it relies on the unpredictable insight of individuals.

Because of the variety of the world, let alone the universe, science is a subject of immense scope. If a different model were required to explain every observation, then we would have gained little through making them. Fortunately that is not how things are; indeed we find that a quite small number of general principles underlies all successful theories of nature. Development of theories along these lines comes itself from a powerful general principle called "Ockham's razor"— simple general explanations are always to be preferred to complicated special ones. These few general principles are so common and so powerful that we sometimes call them "laws"—Newton's laws of motion for example—although sometimes they are called "principles" or even, when expressed in mathematical form, "equations". We shall meet only a few of these powerful principles in this book, but I hope that you will appreciate their importance.

Analytic and Holistic Science

It is fashionable in some quarters to criticise science in the mistaken belief that it is always "reductionist" or "analytical", by which the critics imply that scientists always reduce complicated things to their simplest parts in order to understand them. They contrast this with various "new age" approaches that they regard as "holistic" because they do not try to understand things from a basic viewpoint but rather treat a rather large system as an entity in itself.

It is true, of course, that many of the major triumphs of human understanding have come from an analytical approach. We have come to understand how the stars and planets move through the heavens, how all matter is made up from molecules whose properties can be investigated, what is the ultimate structure of atoms, how plants reproduce by pollination, and so on. We have also come to understand how the behaviour of all these simple entities is controlled by just a few very general principles that can be relied upon to describe with great accuracy what will happen in a huge variety of circumstances.

But this is only one face of science—the other is concerned with interactions and systems and complexity. We know that the behaviour of a complex system, explicable though it may ultimately be in terms of the interactions between its individual components, is often much better considered from a much broader and more "holistic" viewpoint. And this knowledge is not new—it has been applied to plants and animals for centuries, it was applied to whole ecosystems by Darwin in the middle of last century, it was given careful mathematical formulation by Boltzmann and Gibbs more than one hundred years ago. Certainly we keep discovering more about the complexity of large systems, as exemplified by catastrophe theory, deterministic chaos and fractal geometry, but all these developments are firmly grounded on a basic scientific understanding and have precise mathematical formulations.

Modern science is both analytical and holistic at the same time, and has been for centuries. Any "whole" is more than the sum of its parts, and modern science shows us how to understand that whole.

Women and Science

As in nearly all fields of human endeavour, social customs have inhibited the participation of women in science, and it is only recently that these barriers are being broken down. Of course there have been distinguished women scientists for more than a hundred years—the very first computer programmer was a woman, women have won Nobel prizes in most fields of science, including physics and chemistry (Marie Sklodowska Curie won Nobel prizes in both), and steadily more and more women are entering professions in science and engineering.

Of course there are formidable problems, some of them biological and some social. Science, unlike literature, is not something that is easily pursued from a home environment, and the challenges of combining family life with the demands of a laboratory-based science are difficult indeed. Women must often make a difficult choice. But where women have devoted their time and energy to science, the evidence shows that they are every bit as good as men.

Some writers on women's issues maintain that science is somehow "masculine" in philosophy and that there is a suppressed feminine side to science. They imply that "women's science" would be somehow different from "men's science". They draw their support from writers such as Thomas Kuhn who believe that science is somehow "culturally determined". Since these writers generally have no first-hand experience of high-level science, they do not give evidence for these proposals or explain how reliable knowledge, which is the essence of science, can have two different and conflicting expressions. When it comes to choosing what science to support, or which possible applications of science to pursue, of course, there is no argument. Women might well, on balance, choose different priorities from men, although there is often no agreement among either group. Politicians, who control the purse strings, and in some countries at some times even more direct weapons, can of course have an immense influence on the pursuit of science.

The Applications of Science

There is a rather fundamental distinction between science and its applications, or between science and technology. Mind you, not all modern technology has been developed from modern science, though much of it has. Notable exceptions are the ancient technologies of ceramics and metallurgy, both of which pre-date any sort of applications of scientific results. The important thing is that the whole store of scientific knowledge is available for use at a practical level. Some of this knowledge has, at present, no conceivable practical use, but most of it serves as a basis for the understanding of the way in which useful machines and processes operate, and points the way towards improvements. Many modern technologies would have been impossible to develop without the application of scientific results that initially appeared to have no practical value—x-rays, radio, transistors, lasers, nuclear power generation, magnetic resonance imaging in medicine ... The list is very long indeed.

Technology is about reliable ways of doing things, in the same way that science is about reliable ways of knowing things—an engineer learns how to accomplish things, while a scientist learns how to ask and answer questions. Both are needed by a modern society. Technology has its basic principles of design that can mostly be traced back to Newton's theory of mechanics, although there are many new practical design rules that have been developed for specific purposes. Many of these design rules are written down quite formally, and many are enshrined in legislation. Doctors have handbooks of available drugs and guidelines for their use, civil engineers have tables of material strengths and allowable loads on various types of girders, clothing designers know the behaviour of natural and synthetic fabrics, and so on. What this means is that there is generally more than one possible way of doing nearly anything we may wish, but the challenge comes in finding out how to do it most elegantly, efficiently and economically.

Because science and technology between them give immense power to humans, there is an important ethical dimension about their application. Science tells us what is possible, technology tells us what is practicable—other considerations must tell us what *should* be done. This ethical dimension is not itself a part of science or technology, but is a legitimate concern for those who study these subjects. Indeed, because most such ethical decisions will be made by lawyers, politicians, accountants and business managers, it is vitally important that these people should learn about the inevitable consequences of particular courses of action. In our brief presentation here, most of our ethical concern will be focused on the environment, for the questions and alternatives are clear and the decisions pressing, but equally important considerations arise in more social spheres and need to be considered of prime importance in social science courses.

2

Getting the Facts

"Science is built up with facts, as a house is with stones. But a collection of facts is no more a science than a heap of stones is a house." Jules Henri Poincaré

Science makes progress by finding answers to important questions about things in the universe, and there are two steps to this process. The first one is to learn how to ask the right questions. Suppose we want to find a cure for cancer, then it is of little use simply to ask "How can we cure cancer?" The first question should be something more like "What is cancer?" and then, when we find that it is caused by cells multiplying in an uncontrolled fashion, we must ask "How do cells multiply?" and then "What factors can influence their growth?" and so on. We must build up a pyramid of understanding before we can hope to find the answer to the question at its apex.

When someone wants to become a scientist, there are several steps to the training process. One of the first is to learn to use our senses—to observe what happens in nature, to do this reliably, and to make a proper record so that we do not have to rely upon memory. If we see something we don't understand we can then either try to work out what is happening for ourselves, or else discover whether someone else has already found the answer. The first approach is good training, but not economical in the longer run, and we can dispose of many questions by consulting the appropriate books. Of course, just because something has been written in a book does not mean that it is necessarily correct, but it is fairly easy to detect the charlatans.

The whole body of scientific knowledge is written down in books and journals in libraries around the world, and is a priceless heritage that has been built up over more than 2000 years, though most of it has been discovered in the past

150 years or so. Most new understanding is built on this foundation, so that science grows like a pyramid on a firm base. As we discussed in Chapter 1, this foundation of knowledge is not immutable "truth" but consists rather of reliable models. To understand the most recent scientific developments near the top of the pyramid, however, we must understand the foundations on which they rest, and scientists, particularly those who want to go on into research rather than to concentrate on applying known science, must spend many years becoming familiar with their own edge of the pyramid.

Research is the process of finding new understanding, and begins with asking the right questions. The second part of the process is then that of progressively finding answers to the questions that have been asked. This generally involves conducting careful experiments and measurements under conditions that can be easily controlled and reproduced. Without getting the facts in this way we can be led completely astray by imagination or by outside influences.

Only a few scientist spend their time working to establish new fundamental knowledge of nature. Most are engaged in applying what is already known to the solution of new problems, many of them of practical importance. There is not a great deal to distinguish these two kind of activity—it is not so much the sort of problem that one tries to solve as the reason for trying to solve it—and one goes about finding the solution in just the same way. One tries to break down a big and complex question into a set of simpler questions to which one can more easily find the answers and, having done that, one reassembles the knowledge. Usually the reassembly is not simply a matter of adding together the partial solutions—in most cases the whole picture is more than just the sum of its parts, because the complex interactions between those parts are important.

Experiments in Physical Science

Experiment is simplest in concept in the physical sciences, though it often involves much more complex equipment than in the biological sciences. The reason for the simplicity of concept is that the experiments are generally designed to answer very definite questions about atoms or electrons or forces, and these things are very nice and repeatable. Every oxygen atom is exactly the same as every other oxygen atom, every electron is exactly the same as every other electron, so that if the experiment is done carefully enough we should get a perfectly clear and repeatable result.

The problem in the physical sciences is that all the easy experimental measurements have already been done long ago and the answers are well known.

Every decade or two a new field opens up with a lot of new and definitive questions to be answered, and then there is a bonanza of new experimental results, but afterwards there is a long process of tidying up the loose ends and of making sure that everything fits together.

To do these experiments, physicists in particular have continuously developed new experimental equipment that is then found to have uses in a whole lot of other fields of science. Early studies of the behaviour of electrons led directly to the development of the cathode ray tube, which became the picture tube of television. X-rays were discovered while Roentgen was searching for answers to questions about atoms and electricity, but are now indispensable to medical diagnosis. Other medical techniques such as ultrasound scans, magnetic resonance imaging, and positron emission tomography were developed in other studies that had no relation to medical science. Lasers, now so common in CD players and supermarket checkouts, were similarly developed from basic studies of the physics of atoms and molecules. There is no need to prolong the list, though this could easily be done.

Fields of research keep opening up in surprising ways as new questions open new windows on what is going on in the universe. The new areas of chaotic dynamics and fractal geometry are recent examples. The study of some of these areas is inexpensive, and the main requirements are time and intellectual effort, but in some areas of physics the limitation is the expense of the equipment. An example is in the study of the fundamental particles from which everything is made. At every stage when a "fundamental" particle has been identified, the question is whether it is really fundamental or whether it is made up of still smaller and "more fundamental" particles. To answer these questions requires huge accelerators to generate or split apart the particles, and this costs more and more money. The most recent accelerators are so expensive that most of Europe has combined its efforts to build one, and the United States has decided that it is too costly to build one of its own. The cost is, to be sure, much less than that of a week of a quite minor war, but we seem likely to have nearly reached the end of the questions that can be answered in this particular direction.

While this conclusion is disappointing to those whose interests lie in this field, it simply means that science must push forward in other directions. To many scientists the end of this particular road appears as a blessing in disguise, for it means that more attention can now be given to other equally interesting questions that lie closer to hand and that may, in the end, prove to be better worth pursuing.

Experiments in Biological Science

Biological science differs from physical science in several important ways. The first is that, while in physics and chemistry all oxygen atoms are identical, for example, in biology all nominally identical individuals, even at the single-cell level, are subtly different because of their history and present energy state. It is generally therefore not possible to reach conclusions that are quite so definite in biology as it is in physical science. The same goes for repeatability of experiments. In nearly all cases, therefore, one deals with statistical probabilities rather than certainties.

The second difference is that complexity is the very essence of biological science. It is still important to analyse problems at the simplest level first, but the interactions between the elements are all-important. The phenomenon of life itself arises from complex interactions between simple systems and, at a fundamental level, is not yet more than partially understood.

The third difference is that ethical questions inevitable enter into biology. We need feel no compunction about combining a carbon atom with two oxygen atoms or about converting chemical energy to electrical energy, but the common bond of life must make us feel concern for all living things. In practice the strength of our concern, even among those who feel most strongly on the matter, is nicely graded by species and size—small life forms that are biologically remote from humans are treated as essentially inanimate from a moral point of view—but a proper concern pervades all of biological science.

Biological science, then, advances upon two broad fronts. At one level, physiologists study whole animals and whole plants as living systems, while at another level other scientists work on single nerve cells or on the action of particular biological molecules within those and other cells. There is even the science of ecology, which transcends individual plants and animals and considers the functioning of whole populations living together in dynamic equilibrium. In all cases, however, the essence of research is first to find the facts and then to try to interpret them. The traditional tools of biology were simply the unaided senses, but these have been supplemented over the years by tools developed by those studying the physical sciences. First came microscopes, then electron microscopes; first simple chemical stains, then radioactive tracers; first simple chemistry, then elaborate analysis of proteins by x-ray and neutron diffraction supplemented by computer imaging.

Our understanding of both basic phenomena and complex interactions in the biological sciences has advanced at an enormous pace in the past forty years, but the variety of living things is so great that an immense amount more remains to be investigated.

Where Experiment is Impossible

While all of science rests upon factual evidence gleaned from nature, and most such evidence is the result of careful laboratory experiments, there are some areas of science where such experiment is not possible. These areas basically concern the past—how the universe got to be the way it is now. By the very nature of things we can only do our best to build up a self-consistent story that accords with all the evidence we can discover; we cannot travel back in time and watch it happening.

These historical questions arise in both the physical and the biological sciences. We would like to know how the universe began, how the stars and planets formed, and how the earth evolved to its present physical state. We would like to know how life began, how species differentiated, and particularly how and when humans came to be.

Of course there is factual evidence concerning all these things left all around us. We can measure something about the large-scale structure of the universe, we can see stars in various states of development, we can find fossils of early life forms embedded in the old rocks of the earth, and we can try to interpret this into a consistent account. The rules for constructing such an account are simple and hard to disagree with. First of all, the account must be consistent with what we can now observe around us—the stars, the planets, the rocks, the fossils, the current variety of life. Secondly, the mechanisms invoked must be consistent with what we know to be physically possible at the present time. This is a weaker condition, since we might concede that some physical constants, for example, change slowly with time. Thirdly, the account must not invoke any special "once only" mechanisms if an equally satisfactory account can be devised without such a stratagem.

Scientists have been able to devise remarkably satisfactory accounts of the way things got to be as they are. Such theories account fairly satisfactorily for the origin and present behaviour of the universe, for the origin of the chemical elements, for the origin and present structure of the Earth, for the evolution of the present wide variety of life forms, and for the emergence of humankind. The one major gap is an account of the origin of life, and here the problem is that the phenomenon of life itself is still not adequately well understood.

The fact that these accounts all seem fairly satisfactory does not, of course, mean that they are necessarily correct even in the large, let alone in detail. For the most part the mechanisms invoked can be checked and shown to be reasonable in the timescales available, and predictions can be made about missing evidence that could give added support to the theories, but of course direct experiment or observation is not possible. There is room for alternative theories, but before any such theory can be given serious consideration it has to be worked out in the same sort of quantitative detail as the present theories, and it must submit itself to the same queries about its reasonableness and its conformity to the "rules of the game" of explanation. In later chapters we shall return to look at the generally accepted views on these important questions.

3

Change and Energy

"Everything that happens, happens as it should, and if you observe carefully you will find this to be so." Marcus Aurelius Antonius (121-180 AD)

It is change that makes life worth living—a world that was fixed like a museum display might be very beautiful, but nothing would ever happen! For this reason it is not surprising that science devotes most of its energies to examining things that happen, rather than just things that exist.

Many important changes have repeating patterns, which makes them simpler to describe and think about. The sun rises every morning, the moon goes through its phases about once a month, the seasons repeat every year and the sun moves across the background of stars to return to the same place. Even the pattern of birth, growth and death, though not exactly the same for everyone, has a comforting regularity about it. Observing and classifying these patterns is an important preliminary part of science, but the real advance in understanding comes when we can explain how the patterns come about and discover common elements controlling them.

Energy and Heat

One of the most important common elements is the notion of energy. Initially we can just think of energy associated with movement—the faster something is moving, the more energy it has. It is also reasonable to say that the bigger and heavier a thing is, the more energy it has when moving. This is almost all we need to know about energy of motion (or kinetic energy as it is sometimes called after the Greek word for motion) though it turns out that it is the square of the speed (that is the number obtained by multiplying the speed by itself) that is important, rather than just the speed. This means that if we double the speed, we increase the energy by a factor four, so that things moving very quickly have a great deal more energy than we might originally have thought. The energy of motion is thus the mass multiplied by the square of the speed; the useful quantity that is just the mass multiplied by the velocity is called the momentum, but we shall not be meeting it again here.

Long ago people used to think that the sun and the moon were able to keep going round the earth only because they had special gods to push them along. Now we know that we don't need gods for such mundane tasks as this—things just keep on moving unless there is something to stop them. This idea seemed strange to many people when it was put forward as a scientific principle by Isaac Newton some three hundred years ago. After all, we know that we have to keep pushing a bicycle to make it go forward, and that rolling balls simply stop of their own accord, but we shall see that this new idea is actually able to explain all this and a great deal more besides.

Since it is undeniable that things moving on earth do seem to always stop of their own accord, even if this is not true of the sun and the moon, it makes sense to ask what has happened to their energy. Certainly their energy of motion has disappeared, and if that is all we could say then we would not be much further ahead. A little observation helps us out here. Suppose we stop a spinning bicycle wheel by pressing one finger lightly against it, then we find that our finger gets hot—perhaps even painfully hot. This suggests that destroying the energy of motion has somehow produced heat. Since we live in a technological world, we probably also know that a steam engine runs by burning coal, which simply produces heat, yet the mechanism of the steam engine can somehow turn this heat into motion of its wheels. All this suggests that there is perhaps a close connection between energy of motion and heat, and we can reinforce this by calling the heat "heat energy". Perhaps instead of the energy of motion simply being destroyed when something stops, it is converted into heat energy—if we are not actually holding a finger, or a more efficient brake, against a wheel to stop it, then it is actually rubbing against its axle bearings and also, in some way, against the air. We are not usually aware of this heat energy released to the surroundings because a very little heat energy corresponds to a great deal of mechanical energy—if we drop something from a building say 100 metres high and stop it at the bottom in such a way that all its energy of motion is converted to heat in the body itself, then its temperature will rise by only about one degree.

Careful measurements about 150 years ago by the British physicist James

Change and Energy

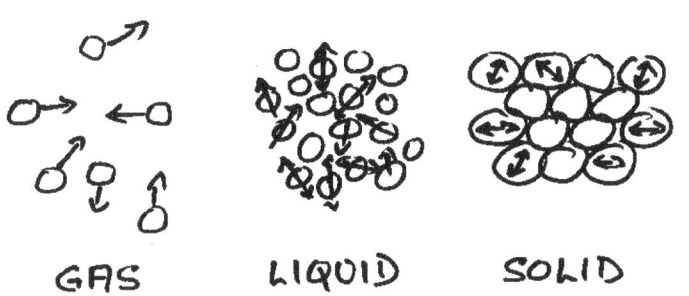

In a gas the molecules are far apart and move around randomly; in a liquid the molecules are nearly touching and both vibrate and move around; in a solid the molecules are in fixed positions and can only vibrate.

Joule showed that this all fits together exactly—the disappearance of a given amount of energy of motion always leads to exactly the same amount of heat energy. This leads us to the conclusion that, despite their very different apparent properties, energy of motion and heat energy are somehow different aspects of the same thing, which we just call "energy". Actually it is quite easy now to see how this comes about, once we know a bit more about heat. Let us look at this in a little more detail, because it will make some of the later things we want to discuss seem more believable. What, then, is heat energy?

We now know that everything is made up of atoms, which are extremely small and slightly soft spheres, often grouped together into tight clusters called molecules. In a gas the molecules are far apart and moving around randomly—which is why it is so easy to move through the air by pushing the molecules out of the way. In a liquid the molecules are packed together so that they are all nearly touching, but they still move around. It is harder to move through water than through air, but still not too hard. In a solid the molecules are all packed fairly tightly together and do not change places, so that the solid can maintain its shape and we have no hope of walking through it. What happens when we heat a solid? However we do it, the molecules begin to vibrate backwards and forwards about their proper positions until they break free and the solid melts into a liquid. If we go on heating, the motion of the molecules becomes even more vigorous and they turn into a gas as the liquid boils. It is thus pretty clear that heat is associated with motion on an atomic scale, so that it is not at all surprising that the ordered motion of a bicycle wheel can be converted into disordered heat motion in a finger tip bringing it to rest. It is much less obvious how we go in the opposite direction by inventing a steam engine, but the Englishmen Thomas Savery (1698), Thomas Newcomen (1712) and James Watt (1764) did just that before the whole matter was understood at all.

Because of Joule's importance in the history of our understanding of heat energy, his name, spelt as joule with a small j and abbreviated J, is given to the unit of energy, whether it is mechanical energy or heat energy or any other kind of energy. We are accustomed, for example, to specifying the energy value of food in kilojoules (1 kJ = 1000 J). Another useful unit to remember is the watt, which is a unit of power corresponding to an energy consumption or output of 1 joule per second. Electrical appliances generally have their input powers expressed in watts.

Just this little amount of scientific principle allows us to understand a great deal about what is going on in the world. Ordered energy of motion is constantly being converted into disordered heat energy as things slow down, but none of the energy is actually lost. Energy of motion can also be passed on from one body to another when they collide, and again we might expect some of it to be converted into disordered heat energy in the process. Heat energy itself can be passed on from one body to another, always going from the hotter body to the cooler one. No energy is really lost, if we keep account of both mechanical energy and heat energy, but it always degrades to less useful forms of heat energy at lower temperatures.

Stored Energy

But how do we go in the opposite direction? How can we wind something up that is continually running down? To understand this, we need to consider some other forms of stored energy. The simplest is the sort of energy stored in a stretched spring. We can put the energy into the spring by doing physical work on it with our muscles, and the stored energy the depends on how hard the spring was to stretch and by how much we stretched it. The spring will hold this stored energy for a very long time, and it can then be released to produce energy of motion, for example by running a clockwork toy.

Just as exerting our muscles to stretch the spring stored energy in it, so exerting our muscles to lift something also stores energy, called in this case energy of position or gravitational potential energy. If we carry a lot of heavy objects

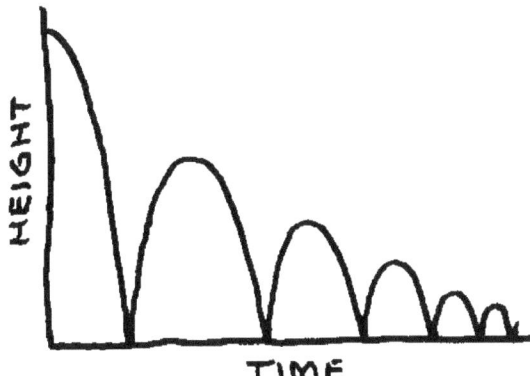

The height of a bouncing ball can be plotted as a function of time. Potential energy (energy of position) is proportional to distance above the ground, and kinetic energy (energy of motion) depends on the speed. The sum of these two kinds of energy remains nearly constant, but decreases slowly because some energy is lost as heat each time the ball bounces.

upstairs, then we would all agree that we have done work and, in a scientific sense doing work means transferring energy. We have transferred energy from our muscles to the objects we lifted. This energy can be converted to energy of motion—in a non-useful way—simply by letting the objects fall. The motion energy is then degraded to heat energy and sound energy as they rattle around on the floor. Much more usefully, water that is stored in dams high in the mountains can be allowed to flow down through pipes to turn machinery which can either do useful work directly or else generate electricity for use elsewhere. Power generated in this way uses the stored energy very efficiently, the only losses being due to the small amount of friction in the large pipes.

We often see potential energy being transformed to energy of motion and back again in everyday activities. Suppose we throw a ball straight up into the air, then it starts off with a lot of energy of motion and not very much energy of position. As it rises, however, it gains energy of position and, at the same time, loses energy of motion. Finally it stops at the peak of its climb and then begins to fall back again, exchanging energy of position for energy of motion. If it hits the ground and bounces, then there is a sudden change of direction and the process repeats, but the bounce is to a rather smaller height than the

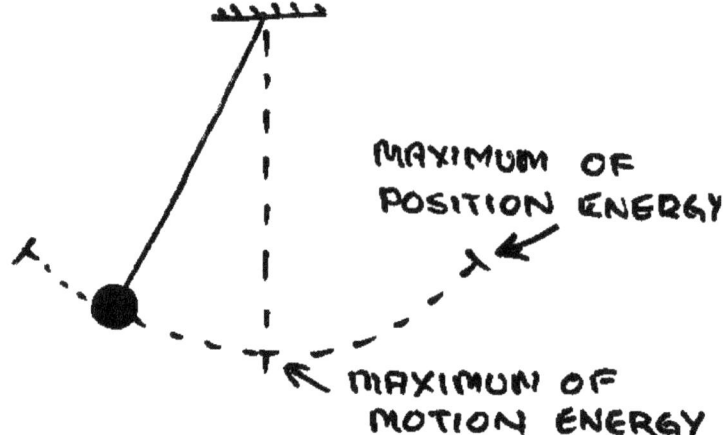

In a pendulum the bob stops at the end of each swing and all the energy is then energy of position. As the bob falls to its midpoint the position energy decreases and the speed, and thus the energy of motion, increases. The sum of the two kinds of energy remains nearly constant but gradually decreases as energy is converted to heat.

original throw because of the energy lost in the rebound from the ground. A "good" ball is one that loses very little energy when it bounces.

Something rather similar happens in pendulum clocks. The pendulum slows down as it rises at each end of its swing, reverses direction smoothly, and then rises up the other side. The only energy lost is that to friction against the air and against its pivots, and this is supplied by the clock spring through a clever mechanism that actually produces a tick as it gives a little push to the pendulum once each swing. The pendulum is a very useful timekeeping device, since it turns out—as Galileo discovered in about 1600 while watching a lamp swing in church—that the time taken for a swing is almost exactly the same whether the swing is a large or a small one. This time does not depend on the weight of the bob of the pendulum either, but only on the length from the pivot to the bob. A pendulum that is one metre long makes very nearly one swing each second, thus repeating its motion about every two seconds.

Energy is stored in another form in fuels such as wood, oil and coal, which can be burnt in air, combining with the oxygen to give useful heat at a very high temperature. Where did this energy come from? Initially from the sun, and we will come back to discuss that later; the answer we need at present is that it was somehow stored in the complex molecules making up the wood, or in the particular atoms making up the coal. It is almost as though the molecules were held together by internal stretched springs, though this is not what really happens. This stored energy, which we may as well call chemical energy, is released again when these complex molecules are burnt to the simple compounds carbon dioxide and water. Because the flames produced by the burning are at a quite high temperature, we can use their heat to turn water into steam, run our steam engine, and create energy of motion. This is not a very efficient process in an ordinary steam engine, because a lot of low-grade heat is lost up the smokestack and from the hot boiler walls. An ordinary steam engine wastes about three quarters of the heat energy of the fuel it burns, and even a very highly efficient large electricity-generating power station wastes about half of the chemical energy in its fuel. What is more, we can prove that such a coal-fired station cannot ever do much better than this—but that is another story.

Food energy is similarly stored chemical energy. The difference is in the way in which it is used, being slowly digested and oxidised at low temperature to carbon dioxide over a period of hours. The chemical energy of the food is made available to the muscles of the body as complex molecules which can then be further broken down as we do muscular work. Food also supplies the complex molecules needed to rebuild the cells of the body and to allow it to grow.

Electric energy is something else we need to consider, because it is so very convenient and useful. Suppose we take two plates of different metals, say copper and zinc, and dip them into salty water. Then this makes a primitive electric battery cell which can produce enough electricity to run a small torch bulb. Where does this energy come from? To answer this we need to ask where the two metals came from, and the answer is that they were recovered from ores mined from underground, and the process of converting these ores to metals used up a great deal of energy. This energy is stored in the pure metals themselves as a form of chemical energy, and can be released as electrical energy when the metals are corroded by the salt solution back to compounds rather like those from which they started. Electrical energy is hardly ever stored directly, except in the capacitors that operate things such as camera flash-lamps. All electric batteries actually store chemical energy.

As well as being made from chemical energy in batteries, electrical energy can be made from mechanical energy in the electric generator invented by Michael Faraday in 1831. The source of the mechanical energy to run the generator can be steam power, generated by burning coal, by nuclear fission, or by solar heating, or can be direct mechanical power from hydroelectric storage reservoirs or windmills. The great usefulness of electrical energy comes from the fact that it can be transported over long distances simply by using metal wires, and can be easily converted to other forms of energy at the receiving end. Electric motors convert electrical energy to energy of motion with nearly ideal efficiency; electric heaters do the same for the conversion to heat energy. And we all know how easily electrical energy can be controlled with switches. Without it our modern electronic society would be impossible. Its one disadvantage is that it cannot easily be stored directly but must be converted to some other energy form first.

Since nuclear energy was mentioned above, we should say a few words about it. Just as molecules have chemical energy stored in the bonds between their atoms, so the nuclei at the centres of those atoms have nuclear energy stored between the protons and neutrons that make them up. Very large atoms decay radioactively to more stable forms, releasing assorted particles together with heat energy, and this provides some of the energy that keeps the earth's core molten. An ordinary nuclear reactor speeds up this process by packing together fuel rods of a particular isotope of uranium in such a way that their atomic nuclei break down to stable smaller nuclei and release great amounts of energy. The fission of one atomic nucleus releases something like one million times as much energy as the chemical burning of one molecule, so that the process is immensely efficient. There are, however, problems with nuclear energy, quite apart from those presented by potential terrorists. Accidents in which radioactive material is released into the surrounding air can happen and are potentially very dangerous. The spent fuel from a reactor is itself also radioactive and must be held in some safe place for as much as several hundred years before its radioactivity is largely exhausted. On the other hand, everything in life poses dangers in return for advantages, and the alternative coal-burning or hydroelectric plants have their own environmental problems.

Solar energy itself derives from nuclear reactions, largely a fusion process in which hydrogen atoms are converted to helium, but this goes on at a distance from earth that makes it quite safe. Ultimately we may use solar energy for all our requirements, but for the moment it is too expensive by about a factor ten to be economically competitive except in remote locations. Wind power, wave and tidal power, and hydroelectric power all ultimately derive their energy from the sun (or the moon in the case of tides) and are environmentally attractive. We discuss some of these matters in our chapter on the environment.

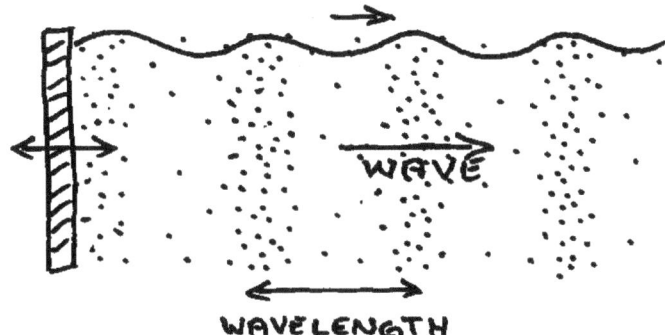

A sound wave is generated by a vibrating object. It is a succession of high and low pressure regions in the air, moving forward at a speed of 340 metres per second. The distance from one high-pressure region to the next is the wavelength.

Light Energy and Sound Energy

There are two other kinds of energy that we need to mention in order to make our survey reasonably complete. These are sound energy and light energy. Both are rather different from the energy types we have discussed before because they are not so clearly associated with material bodies and are rather difficult to think of in a concrete way. Both are, in fact, types of wave energy, and it helps to begin with the more familiar waves on the ocean or on the surface of a pond. It is clear that these waves move with some sort of characteristic speed and that they are also distinguished by what is called their wavelength—the distance from one wave crest to the next. For ripples on a pond the wave speed is about one metre per second and the wavelength is typically a few centimetres, though this depends upon how the wave was excited.

Sound waves are excited when something vibrates very rapidly, such as a sheet of iron that has just been hit with a hammer. The motion of such a solid vibrating object compresses and decompresses the air near its surface several hundred times each second, and these compressions and rarefactions move off through the air as sound waves with a speed of about 340 metres per second. Humans can hear sound with frequencies between about 20 and 20,000 vibrations per second, and thus with wavelengths from about 17 metres down to 17

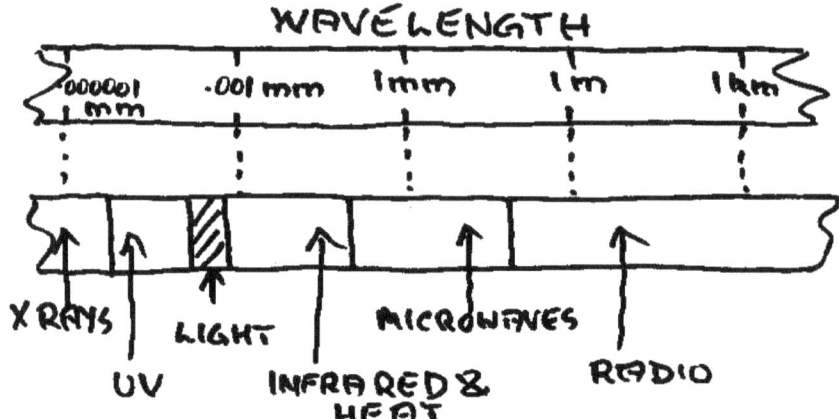

Light is just one small part of the electromagnetic spectrum which ranges from x-rays at its short-wavelength end to radio waves at long wavelengths.

millimetres. Just as waves breaking on a shore carry energy that erodes cliffs and sandbanks, so sound waves carry energy on a much smaller scale. When we have a collision between two objects, as well as each of them retaining some motional energy, there is energy converted to heat and sound.

Light waves are different and altogether more complex. They are not waves in a material medium like the air, but rather waves in electric and magnetic fields. We need not worry too much about this except to note that, since no material medium is required, light waves are able to travel across the nearly complete emptiness of space without losing any energy. Just as our ears can hear sound waves of only a limited range of frequencies or wavelengths, so our eyes are sensitive only to a limited range of the electromagnetic waves that we call light. This visible wavelength range is from about 400 to 700 nanometres (thousand millionths of a metre) and the smallness of the wavelengths involved explains why we can see such fine detail. Electromagnetic waves cover all wavelengths, however, those shorter than the visible range being ultraviolet radiation and x-rays while those that are longer are infrared radiation, microwaves and radio waves. Anything that is hot enough gives out infrared heat radiation and light radiation, so that this particular conversion is very simple.

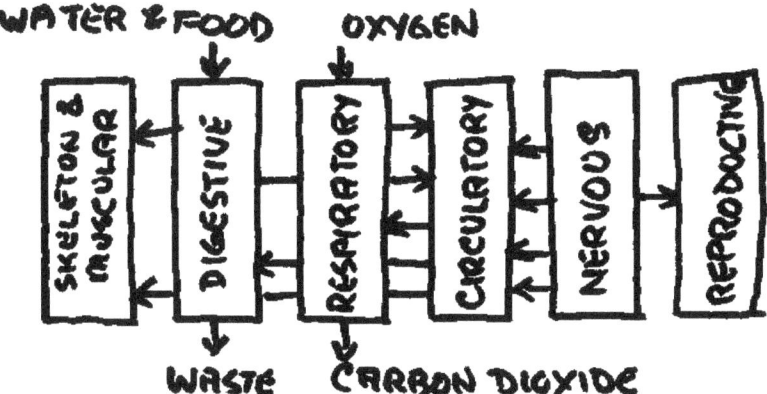

In the operation of any machine or animal, there is some "useful" work done and there is always some wasted energy which usually appears as heat lost to the surroundings. Some systems may store energy for use at a later time, as in the elastic energy of a spring or the food energy stored as fat in an animal.

Wasted Energy

We have seen that energy is neither created not destroyed but, when we consider all types of energy, simply converted from one form to another. This statement is known as the principle of conservation of energy, and it has been an extremely powerful guide in the development of science. It has been tested in countless ways and always found to hold. A colloquial version might be phrased "in energy matters, as in the rest of life, you can't get anything for nothing." Perpetual motion machines are simply impossible.

With this principle in mind, it is interesting to examine some systems to see where the energy goes. If we were to do this quantitatively, we would be surprised, in most cases, to find how much is wasted. To do this would take us too far into matters of detail, but a few examples might make the point.

A motor car wastes a great deal of the energy in the petrol consumed because the fuel-burning cycle of the engine cannot be very efficient. This waste energy, perhaps 60 percent of the chemical energy in the fuel, is lost as heat from the engine and exhaust. A very small amount is lost as sound energy and light, if the headlights are on. The car then has to overcome internal friction in its moving parts, it has to stir up the air to get through, and its tyres lose energy through road friction. All this appears as heat and noise. When the journey is over, there may be some stored energy in the car and its passengers, if it is now at the top of a hill, and this can be recovered by coasting down the hill and building up speed. If the journey has been on the flat, then there is no residual energy and everything has been lost as heat and noise. Of course, if the occupants of the car had walked to their destination, they also would have consumed food energy to do so. In this case it is hard to define total efficiency, because all work is done against frictional and other losses.

A radio set is another example of interest. The input is the electricity from the power mains, perhaps 100 watts of power (100 joules per second). There is a tiny input from the radio signal, but it is so small that we need not take it into account as more than a controlling agency. The radio puts out useful energy as sound and it turns out that, even for quite a loud radio, the output sound power is less than one tenth of a watt. More than 99 percent of the input energy has gone into running the components of the radio and has finally appeared as heat, with a very little appearing as light from the illuminated dials.

4

Hearing and Seeing

"And God said, Let there be light: and there was light." Genesis 1:3

Most of our information about the world we live in is gained through our five senses—touch, smell, taste, hearing and sight—but the most important of these are the abilities of our ears and eyes. We have already read a little about sound waves and light waves as carriers of energy in Chapter 1; it is now time to look at sound and light in rather more detail.

Sound and Hearing

Sound is produced whenever something vibrates in air. In fact sound is produced under water in the same way, but we will not have time to consider this here. There are three things that characterise a steady sound—its loudness, which is related to the amount of energy in the wave, its pitch, which is related to the frequency of vibration of the object generating the sound, and its tone quality (or timbre as musicians call it) which is related to details of the vibration. If the sound is not steady, as for example a hammer blow or the note from a guitar string, then we need to say how these three properties of the sound vary during its duration. This can all become very complicated and leads into subjects such as music and phonetics, but a few more simple ideas will suffice to give us a reasonable understanding.

First—loudness. Fairly clearly, the harder we hit an object the more vigorous is its vibration and the louder the sound it produces. This seems sensible, since the motion of the object produces pressure waves in the air and these pressure waves cause our eardrums to vibrate, so that these should all increase together. A very small object, however, is very inefficient in compressing the air when it vibrates, so that it produces only a soft sound even for quite large vibrations—try this by plucking a stretched rubber band. To produce a louder sound we need first to transfer energy from the small object to something larger that can then transfer it to the air. The guitar and the violin are excellent examples—the vibrating strings produce little sound by themselves, but they can pass on vibration energy to the large light wooden body of the instrument which then radiates it efficiently.

The human ear is a remarkable instrument in the range of sound intensities that it can accommodate. From the softest sound that we can hear, such as a distant whistle on a windless night in the country, to the loudest sound that we can tolerate—something a little more than a loud rock band—is a factor of one million million in sound energy. Our ears have not evolved to cope with sounds at the loud end of this range, which do not occur in nature, for more than a few moments without damage, however. Long-term listening to such loud music gradually destroys the nerve cells of the ear, so that people who do this become increasingly deaf as they grow older.

Next—pitch. The rate at which an object vibrates when it is disturbed depends upon its mass (or weight) and its stiffness. The stiffness of a rubber band or of a string can be increased by stretching it tightly, and as we do so the pitch of the sound will rise. This is how we tune guitars and violins. Another way to alter the pitch is to keep the tension the same and to shorten the length of the string that is allowed to vibrate, which is how we play different notes on these instruments. If you look inside a piano you will see that the strings for the high notes are short, while those for the low notes are long and wrapped with copper wire to make them heavy.

If we start with something that is already stiff, however, such as a metal rod, then the only way we can change its pitch is to change its dimensions. A very short rod is much harder to bend than a long one of the same thickness and, even though it also weighs less, the increase in stiffness wins and the pitch rises as we shorten the rod. If we have several metal rods of the same length but different thicknesses the thin ones will be easier to bend than the thick ones, so their pitch will be lower. These principles are used to design sets of bells and other metal instruments on which we wish to play tunes.

Long columns of air enclosed in tubes also have a sort of elastic stiffness, since they resist being compressed, and can be made to vibrate lengthwise, rather than from side to side, by buzzing lips (in trumpets), vibrating reeds (in clarinets) or vibrating air jets (in flutes). Once again, as we make the air column shorter by closing valves (in trumpets), by sliding sections of the pipe (in trombones) or by opening holes in the side of the tube (in clarinets and

Hearing and Seeing 27

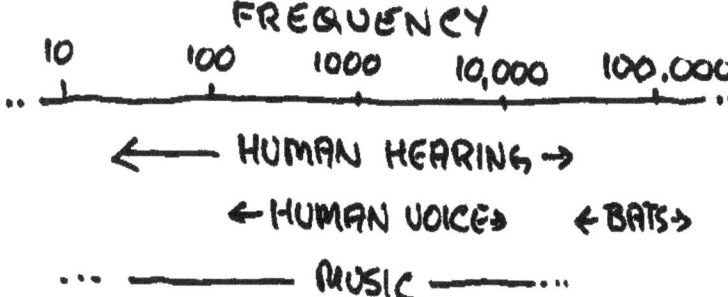

The range of human hearing extends from about 20 Hz to 20,000Hz. The fundamental pitches of human voices range from about 100 to 600 Hz, though these sounds are accompanied by overtones that extend up to about 5000Hz. Musical instruments cover most of the range of human hearing.

flutes), the pitch of the note being played gets higher.

Pitch is what is called a subjective property of a sound—we need a human to tell us about it, rather than being able to measure it directly. The thing that we can measure is the frequency of the sound, which is the number of vibrations per second (we use the unit hertz, abbreviated Hz, for this measure, after the German physicist Heinrich Hertz who first produced radio waves in 1888). Again the ear has a very wide frequency range from about 20 vibrations per second to about 20,000 vibrations per second, although it is most sensitive in the range from about 200 to 5,000 vibrations per second which is the range in which most of the energy in human speech is concentrated. A doubling of frequency corresponds to an octave on the piano, so that the basic pitches of a piano go from about 16 Hz to about 3,500 Hz, though the lower frequencies in particular are accompanied by many higher "overtone" frequencies as we discuss next. Most animals have hearing ranges that are not very different from those of humans, except for bats which are able to hear extremely high-pitched sounds which they use as an echo-guidance system.

Finally then—tone quality. Even simple sounds can be bright or dull, smooth or discordant, and this is because they do not consist of a single frequency but rather of a whole host of frequencies called overtones. If these overtone frequencies are simply related, as they are in a plucked or bowed string, in a simple air column, or in a carefully made bell, then the sound is pleasant and concordant. If the overtone frequencies are all over the place, as they will be in most cases, then we either hear a dull thump as in a drum, a metallic clang as when we hammer a metal sheet, or an unpleasant scream as in a circular saw. On a gentler basis, the difference between vowel sounds in human speech—oh, ah, ee, etc—arises from the different energies of the overtones, as determined by the shape of our mouth and lips when we speak.

Human voices, the barks of dogs and the songs of birds are all produced by the interruption of a flow of air out of the lungs by pieces of vibrating membrane—the vocal folds in humans. In children these folds can vibrate somewhere between about 300 and 800 times a second, and we control this rate when we sing. Women's voices are a little bit lower, about 200 to 600 Hz, while men's voices are pitched nearly an octave lower at about 100 to 300 Hz. Of course, trained singers have much wider ranges than this. These are just the fundamental frequencies of speech—as we noted above, they are accompanied by overtones that extend up to about 5000 Hz.

There is clearly a great deal more we could say about sound and its importance to us in speech and in music, and we could go on to talk about the use of ultrasound—very high-frequency sound that will penetrate human tissue without doing damage—to examine the insides of our bodies, but here we must leave the subject for now.

Light and Seeing

As we saw in Chapter 3, light is a form of electromagnetic energy that is emitted from hot bodies such as the filaments of light globes. The temperature of these filaments is about 3500°C and the light is somewhat yellowish compared with that of the sun, the surface of which is at a temperature of about 6000°C. If a piece of metal is only "red-hot" then its temperature is probably only about 1000°C. Light can also be produced in rather more complex ways from glowing gases in neon signs, from special phosphor materials on the tubes of fluorescent lights and the screens of television sets, and in very subtle ways from lasers.

The eye is a very sensitive detector of light and can respond to a great range of intensities before we are dazzled, but it only detects a very small part of the total range of electromagnetic radiation, which extends from radio waves through to x-rays as we discussed in Chapter 3. Indeed the human visual range is only from wavelengths of about 400 thousand-millionths of a metre, which we

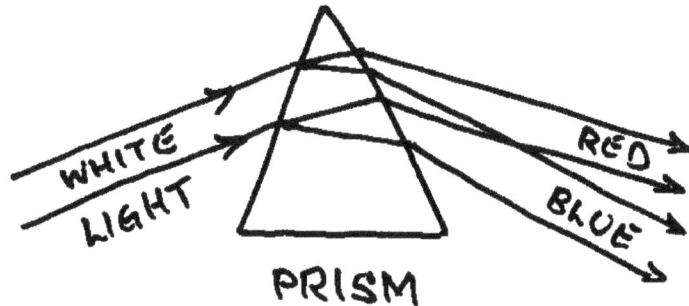

When white light is passed through a glass prism, it is split into all the colours of the spectrum—red, orange, yellow, green, blue, indigo, violet.

interpret as the colour violet, to 700 thousand-millionths of a metre, which gives the sensation of red. (One thousand-millionth of a metre, incidentally, is called a nanometre for short.) This range of vision has evolved so that our eyes take best advantage of the light from the sun, which is concentrated between these wavelengths, though there is also a good deal of infrared radiation, which we feel as heat, and some ultraviolet. Most of the ultraviolet radiation is stopped by the ozone layer high in the atmosphere, but some gets through to cause sunburn and even skin cancer.

As Newton showed when he passed white light through a glass prism and found all the colours of the rainbow, the sensation of white is a neutral one that we have developed to describe the balance of the light from the sun, while all the colours tell us something about the mixture of wavelengths in the light we are seeing. As for pitch in hearing, this is something subjective and we must learn about it by asking people what they see, or by asking them to match colours. A little information about the eye will help us to understand what happens.

In our eye, light from the scene we are seeing is focussed onto the retina at the back of the eye—we look at the action of lenses in a moment—where it stimulates special nerve cells that carry information to the brain. There are two main types of cells, called rods and cones because of their characteristic shapes. The rods are much more sensitive than the cones and are the means by which we see in very dim light. They do not, however, carry any colour information, so that scenes viewed by moonlight or other dim illumination are

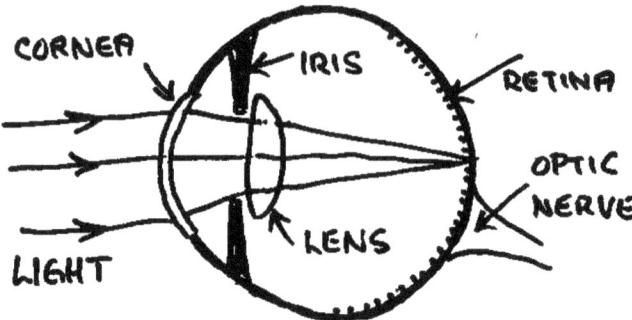

Light entering the eye is focussed onto its back surface (the retina) by the action of the curved front surface of the eye (the cornea) and the internal lens. Sensitive rod and cone cells in the retina are stimulated and convey nerve impulses to the brain for analysis.

basically black-and-white. The cones, on the other hand, have pigment dyes in them of three different colours, so that one family of cones is sensitive to blue light, one family to yellowish green, and one family to reddish yellow. When light falls on the cones, it is the balance between the amount absorbed by each of the three families that determines the colour sensation that we experience.

It is useful to simplify this a little and speak of red, green and blue-sensitive cones, for then we can see easily how colour mixing takes place. On a television screen there are patterns of tiny phosphor dots which give, out red, green and blue light respectively when they are activated by the electron beam in the tube. If just the green dots on the TV screen emit light, then only the green-receptors in the eye are stimulated and we see the colour green. The same happens for red and blue. If both the red and green phosphor dots are activated, then the red and green light stimulates the red and green receptors in the eye in very much the same ratio as they would be stimulated by pure yellow light, and we see colour yellow. Similarly for other mixtures—red and blue dots give a purplish red sensation that we call magenta, while green and blue dots give a light-blue colour called cyan. This process is called additive colour synthesis because we are adding the coloured light produced by the phosphor dots on the television screen. Red, green and blue are the "primary colours" of additive synthesis. If we add together these three primary colours on a TV screen then the result is

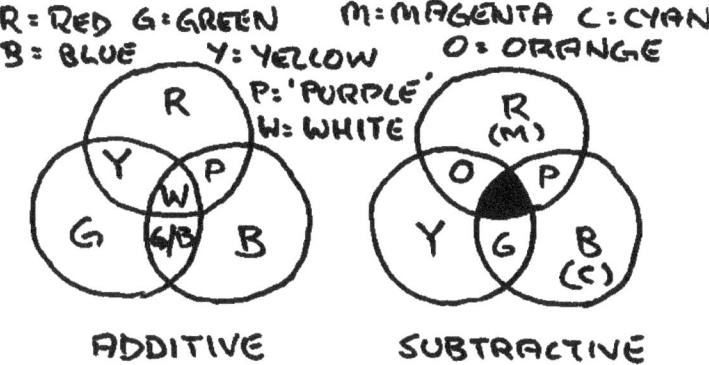

In additive colour synthesis, as in a colour TV, red, green and blue primary lights are added to make the whole range of possible colours. In subtractive colour synthesis, as in painting or printing, yellow, cyan (blue) and magenta (red) dyes blank out parts of the white light reflected by the paper to again give the whole range of colours.

white—you can see this by looking at a TV screen with a magnifying glass.

Anyone who has experimented with painting knows, however, that artists deal with three different primary colours which they call red, yellow and blue, and mix these three to obtain all their other colours. What is the difference? To see the explanation we must think how artists colours work. The artist basically uses a transparent colour painted over a white background, as is easy to see in the case of water-colour painting. White light from the sun then passes through the pigment film, has some of the colour absorbed out of it by the dye, and is reflected back by the white background. The same sort of thing happens with oil paint except that the white material is suspended as powder in the paint itself. The process by which painters produce colours is called subtractive synthesis, since the paint layers, or the mixed paints, take successively more and more colour out of the original white light. The three colours we need to do this might be called "minus-red", which leaves only the green and blue part of the spectrum and gives the light blue called cyan, "minus-green" which leaves red and blue and gives the purplish red magenta, and "minus-blue" which leaves red and green and gives the sensation of yellow, just as we saw above for additive synthesis. Thus to get green by a painter's subtractive-synthesis method, we

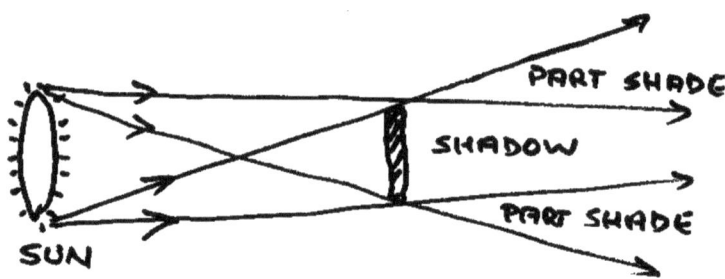

Rays of light from the sun, or from a lamp, travel in straight lines to produce shadows. Because the sun is a disc and not a point, there are regions of partial shadow surrounding the main shadow.

add blue cyan (minus-red) and yellow (minus blue) so that the white light loses both its red and blue and only the green part remains. If we mix together all three artists' primary colours, then all the white light is absorbed and we get black, or at least a dark "dirty" brown.

Lenses and Mirrors

One of the most practically important things about light is that it travels in straight lines without spreading, because of its short wavelength. We can see this easily when we make sharp shadows using a very small light source. Shadows actually have blurred edges because light sources such as the sun are not ideally small, and the light from one part of the sun's disc gets into the shadow from another part.

The reason this is important is that it allows us to make simple optical instruments such as mirrors, and then more complicated instruments such as telescopes and camera lenses. Mirrors are the simplest instruments to think about, so let's examine them first. The basic principle of reflection is that light behaves like a tennis ball thrown against a brick wall—it is reflected at an angle that is the same as the angle at which it hits the wall. This rule is all that we need to understand even the most complicated reflections. The light from a candle held in front of a mirror is reflected so that the light rays all appear to

Hearing and Seeing

come from a phantom candle behind the mirror, and we can extend this picture to explain how we see the whole of the room reflected in the mirror.

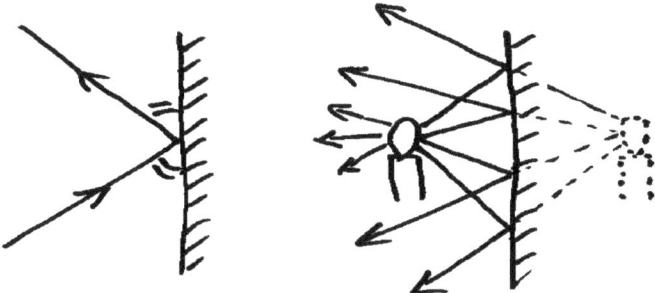

Light travels in straight lines. When it is reflected from a mirror, the angle at which it is reflected is the same as the angle at which it strikes the mirror. This simple rule allows us to understand how images of objects appear to be located behind the mirror, as shown for a candle.

Another important sort of mirror is the curved mirror used in astronomical telescopes or, in a more mundane application, to focus the rays of the sun in a solar cooker. We use a mirror that is part of the surface of a sphere or, more carefully, part of the surface of a paraboloid which is a little different. Because the sun is a very long way away, all its rays are very nearly parallel, and simply by drawing reflections for each ray we find that they all pass through a single point called the focus. If the mirror is part of a telescope, then we can place a photographic plate at its focus and take a photograph of the sun—or more realistically of the stars, for which the principle is the same. If we just want to make a solar cooker or the boiler for a solar power station, then we put a black-coated container at the focus of the mirror, and the sun's radiation heats it up to a very high temperature.

Most simple telescopes and cameras use glass lenses instead of mirrors to focus the light. Their operation is a little more complex, and the light is bent as it passes through the curved glass. Near the centre of the lens the two glass surfaces are nearly parallel and there is very little bending, but away from the centre the light rays are bent more and more so that they all come to a focus at a single point. We can therefore use a lens in very nearly the same way as a mirror.

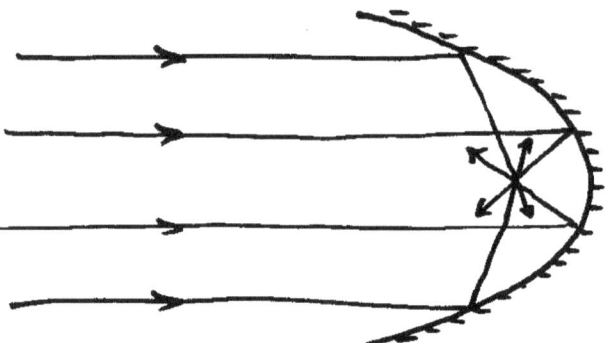

Rays of light from the sun can be focussed by use of a curved mirror. A mirror whose surface is part of a sphere works well, but a mirror with a surface that is paraboloidal, as shown, works even better.

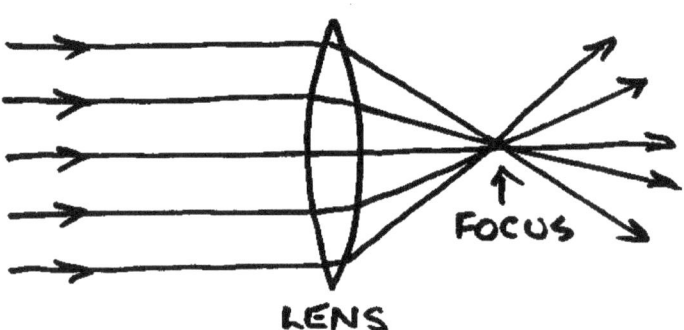

A glass lens can focus the parallel rays of light from the sun down to a point, because rays going through the edges of the lens are bent more than those going through the centre. This principle allows a lens to form an image of a distant object on a sheet of paper placed at its focus, and is the basis of the photographic camera.

Depending upon just how we set them up, lenses and curved mirrors can be used to make things appear either larger or smaller, and we can make this work even better by combining several lenses into a telescope or a microscope. The details of how this is done need not concern us here. It is useful to note, however, that lenses can also be used to correct for defects in the natural lenses of our eyes and make our vision clearer. Some people have naturally good eyesight and do not need glasses, but for others the correcting power of a glass lens makes all the difference between seeing clearly and seeing a badly focused blur.

5

The Universe and Everything

"And he sees the vision splendid of the sunlit plains extended, And at night the wond'rous glory of the everlasting stars." 'Banjo' Paterson "Clancy of the Overflow"

It has been known for more than 2000 years that the Earth is round and rotates on its axis. The Greek philosopher Aristotle (384–332BC) concluded that the Earth is round from its shadow on the moon during an eclipse, and the Greek astronomer Eratosthenes (276–194BC) was later able to measure its diameter by comparing the lengths of the shadows cast by identical vertical poles in two places separated by about 900 kilometres, obtaining a result that was only about 15 percent different from modern measurements. Philosophers had mixed views on the rotation of the Earth about its axis, and here observation was of no help whatever. A model with a fixed Earth and everything, including the distant stars, rotating about it gives exactly the same predictions as does a rotating Earth with fixed distant stars—it is simply a changed point of view. It is only when one tries to construct a model of *why* things should happen this way, based upon a few simple principles, that the rotating-earth model is clearly superior.

Something rather similar happens when we consider the motion of the sun relative to the Earth, or vice-versa. The astronomer Aristachus of Samos (310–230BC) put forward a model for the universe in which the Earth moves in a circular orbit around the sun, very much like our present view. However the alternative picture in which the sun moves around an immovable Earth fixed at the centre of the universe gained popular approval with Greek philosophers who felt that man should be at the centre of all things. This model is generally associated with the name of Claudius Ptolemy of Alexandria (100–178AD).

Remembering our discussion of scientific models in Chapter 1 and the comment above on the rotation of the Earth, it is interesting to note that neither of

The Universe and Everything 37

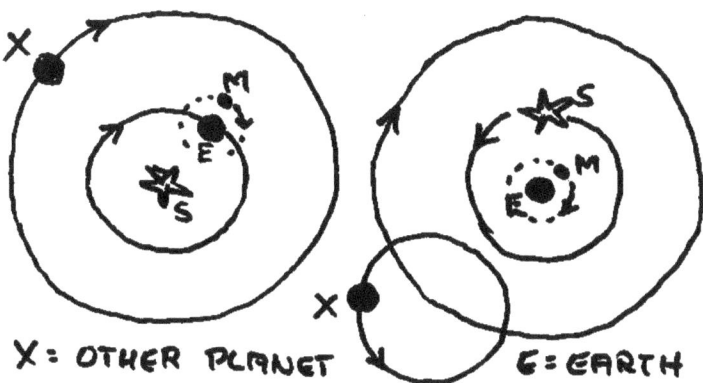

Models of the solar system with the Earth at the centre and with the sun at the centre both give equally good representations of the motion of the planets, but the sun-centred system is much simpler. The Earth-centred system involves compound motions called epicycles.

these models is actually "wrong"—they simply represent different viewpoints. Each is able to describe the motions of the sun, moon and planets equally well, although the Earth-centred model is geometrically more complicated and requires each planet to follow a double motion called an epicycle. However both models run into problems if one tries to make them accurate, because the Greeks sensibly took the orbits of the planets to be circular whereas they are really elliptical.

In the early middle ages a variety of "flat Earth" views deriving from Hebrew ideas were widely accepted in the Christian world, while the more enlightened ideas of the early Greeks were kept alive in the Arab world. These classical ideas were brought back to Western Europe around 600AD, so that the common view of later medieval writers was in favour of the Earth-centred model of Ptolemy. It served for prediction of the positions of the moon and planets, but involved a very complex set of calculations.

The model with the sun at the centre and the Earth rotating about it was resurrected by Nicolaus Copernicus (1473–1543) and finally found to give an elegant and simple account of the solar system by Johannes Kepler (1571–1630) once the old circular orbits were replaced by ellipses. An ellipse, incidentally,

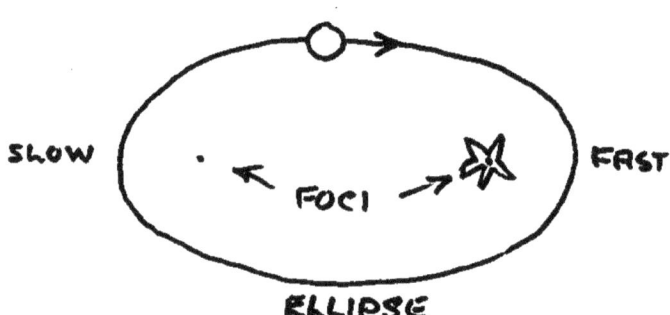

The true motion of planets is in an elliptical path with the sun at one focus. The actual orbits of the planets are, however, much more nearly circular than shown here.

is the shape seen when a circle is viewed at an angle so that it is flattened in one direction. It can be drawn by passing a loop of string around two pins and running a pencil around the orbit while keeping the loop tight. The pins are at the two foci of the ellipse, and in the case of planetary orbits the sun actually lies at one focus.

The importance of the sun-centred model, as popularised particularly by Galileo Galilei (1564–1642), is that it allows not just a description of the motions of the solar system, but also an explanation of those motions in terms of a few simple principles. This was one of the great accomplishments of Isaac Newton (1642–1727). The simple assumption of a gravitational force falling off in strength as the square of the distance was all that was needed, though this proposal took a great leap of scientific imagination, and the working out of the actual laws of motion laid the foundations both of modern science and of the branch of mathematics known as calculus.

The Solar System

The modern picture of the solar system is too familiar to require much discussion. The sun is immensely large, with a mass that is nearly a thousand times the combined masses of all the planets. It produces energy by combining hydrogen atoms to form helium in a nuclear fusion reaction. The central temperature

The Universe and Everything 39

The rotation axis of the Earth is inclined at about 23° to the plane of the Earth's orbit (the ecliptic) and always maintains its direction in space. It is summer in the southern hemisphere when the tilt of the axis exposes it more directly to the sun, as shown.

of the sun is more than a hundred million degrees and the pressure is immense, but at the surface the temperature is only about 6000°C, giving an appearance that we interpret as "white hot". The planets move around the sun, all nearly in the same plane which is called the ecliptic because that is where eclipses can occur, in slightly elliptic orbits, and many of them have several satellites like the moon moving around them in turn. In the case of Saturn, there are thin rings of fragmentary material rotating around the planet as well. The inner planets, Mercury, Venus, Earth and Mars, are all more of less rocky in composition, some of them with cores of molten metal, mostly iron. The Earth and Venus are about the same size, with Mars only one tenth and Mercury only one twentieth of the mass of the Earth. The outer planets except Pluto, in contrast, are huge, with Jupiter the largest being about 300 times the mass of the Earth, and their composition is entirely different, consisting largely of hydrogen and ice. Modern space probes have, of course, given us immensely detailed information about the surfaces of these planets and about their moons.

The axis of the Earth is not exactly perpendicular to its orbit (the plane of the ecliptic), which leads to important effects. The main one is that the sun's apparent motion across the sky varies with the season of the year. In January the north end of the Earth's axis is inclined towards the sun, so that the northern hemisphere gets longer hours of sunlight and warmer temperatures

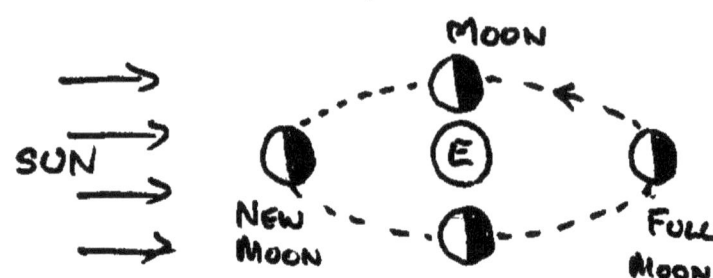

As the moon moves around the Earth in its orbit about once every 30 days, we see a different amount of its surface illuminated by the sun. The moon always keeps the same face directed towards the Earth.

as its summer season. In July, six months later, the north end of the Earth's axis is inclined away from the sun and it is summer in the southern hemisphere. The actual tilt of the axis is about 23° so the sun is directly overhead at midday in some month (actually two different months except at the latitude extremes) at any place with latitude between 23° north and 23° south of the equator, a region known as the tropics. At the north and south poles, the sun moves around at a constant angle of 23° above the horizon in the middle of summer, and sinks to 23° below the horizon in the middle of winter, giving continual darkness.

The phases of the moon are common knowledge and occur because the moon has only its sunward hemisphere illuminated. Because the moon revolves around the Earth once every 27 days and the Earth itself revolves around the sun, we see a complete cycle of changes in the phase of the moon every 30 days. The moon occasionally moves into the Earth's shadow, in which case there is an eclipse of the moon—not a particularly rare event because the Earth's shadow is a good deal larger than the moon. Much more rarely the moon comes directly between the Earth and the sun and we have an eclipse of the sun. As seen from the Earth, the moon covers just about the same angle in the sky as does the sun—about one degree—so that it is extremely rare for the eclipse to be total, and even then it can be seen from only a small part of the earth. At other times the eclipse is partial, with just a fraction of the sun's disc obscured by the

The Milky Way galaxy in which our sun is located is an immense spiral structure. There are hundreds of millions of stars in the Milky Way galaxy, and hundreds of millions of galaxies of similar size in the universe.

moon.

A Modern View of the Universe

The sun is a star of rather average size and we can see many more like it in the night sky. Nearly all the stars we can see belong to an immense flat spiral cloud that is called the Milky Way galaxy which we can see spread out in a bright band across the sky. It contains perhaps one hundred thousand million stars altogether and is immensely large. In fact astronomical distances are so large that astronomers measure them not in kilometres but rather in light years—the distance that light travels in one year at its immense speed of 300 million kilometres per second. On this scale, the sun is about 8 light minutes away from the Earth, our nearest star is about 4 light years away, and the Milky Way galaxy is about 100,000 light years across, with the sun being about 33,000 light years from its centre. The whole galaxy rotates very slowly—about once in a thousand million years.

As if this were not enough, the universe is populated by thousands of millions of galaxies, all of size comparable to the Milky Way. There are two very small ones close to us—the Magellanic Clouds that can be seen just to the side of the Milky Way—but most are immensely far away. To make things even stranger, astronomical measurements show that all these galaxies are moving away from us at immense speed, with the farthest ones moving most rapidly! When we examine these velocities carefully, we find that the behaviour is just as if there was an immense explosion about 15,000 million years ago that blew the universe apart. Naturally the galaxies that are moving fastest have moved away the greatest distance. Even this, however, is not as simple as it sounds. We are not in a special place at the centre of the universe with everything moving away from us, rather the whole universe is expanding uniformly. It is hard to picture this for a real three-dimensional universe with length, breadth and height, but we can get an idea of what is happening for an imaginary two-dimensional universe of galaxies drawn as dots on the surface of a balloon. As the balloon is blown up, the dots get further apart, and those that are more widely separated move apart most rapidly, but there is no centre of expansion within the surface of the balloon—it lies away at the centre of the balloon in the third dimension. Similarly our real universe requires four dimensions to describe properly what is going on, but this is far too complicated to discuss here.

This "big-bang" theory of the origin of the universe appears to describe it origins and evolution very well. The theory is, in fact, very much more complicated than outlined here, and shows also how elementary particles and atoms were produced in the immense temperatures just after the explosion. The theory even predicted the temperature to which the radiation in empty space would have cooled by now—about $-270°C$ or three degrees above absolute zero—and this radiation was later found and measured. It is not really possible to discuss the details without using mathematics and relativistic mechanics, which blends together time and space so that questions such as "What was the universe like before the big bang?" do not have any meaning because there was no such thing as time then. As you may have suspected, of course, the big-bang theory is at present only the best guess that science has made about the origins of the universe; it is by no means absolutely established. There are some other candidate theories, but none seems to be nearly as successful in explaining what can be observed and measured.

As well as ordinary stars like the sun, there are many peculiar objects in the universe. It must suffice to mention a few names with brief descriptions. Quasars are possibly distant galaxies in collision and emitting intense radio waves. Pulsars are small collapsed stars ("neutron stars" because the atoms have all turned into a sort of nuclear soup) rotating very rapidly and emitting a narrow beam of radio waves that flashes past the Earth like a searchlight beam with great regularity every second or so. Black holes are larger stars that have collapsed so far that their intense gravitational field prevents anything from escaping, even their own light.

The Future of Us All

The Earth, which is composed of material left over from the great cloud of dust and gas that collapsed to form the sun, has had a solid crust for perhaps 4500 million years. Life has been around for around on its surface 3000 million years, as we discuss in Chapter 11, and the human race for only about the last one million years, a very tiny part the history. It seems presumptuous, perhaps, to inquire about our future, but naturally this does concern us. As we discuss in a later chapter, the immediate threats to human life are not bound up with the natural evolution of the universe, or even of our planet Earth, but are troubles of our own making. It is interesting, however, to look at the future in broader perspective.

Stars such as our sun do not last for ever—they have only a finite amount of hydrogen fuel, and when that is exhausted something drastic must happen. The death of a star depends upon its size, but involves either expansion to a red-giant size that would nearly engulf the Earth, or else explosive collapse to a white dwarf. In either case, the Earth would not survive, but the timing of this event is some thousands of millions of years into the future, so it should not be of real concern to us.

On an even longer time-scale, the universe may go on expanding for ever, gradually cooling down as its stars become extinct, or it may have a cyclic behaviour so that, after an immensely long time, the expansion stops and reverses, compressing the whole universe back to the immensely hot and compact singularity from which we think it started. Obviously we will not be around to check out either of these possibilities directly, but we can continue to build models, test them against the observations we can make, and then explore their long-term predictions. This is still science, because it relies upon careful mathematical analysis and detailed comparison with experiment, but it begins to answer some of the problems that philosophers used to ask before physics took over their mantle.

6

This Earth of Ours

"At the round Earth's imagined corners, blow your trumpets, angels." John Donne (1572–1631)

The rocks on the Earth's surface have been solid now for something like 4500 million years. How do we know? Well, basically from an examination of the slow decay of naturally radioactive materials that were included in the rocks when they first solidified. Before radioactive dating methods had been established, geologists could make only very rough estimates based on indirect data, but now there are hosts of independent measures with different radioactive atoms that not only allow us to identify the very oldest rocks, but also to put reliable dates on the newer rocks as well. Indeed, related methods based on isotopes of carbon—natural carbon atoms that are chemically identical with ordinary carbon but that have an extra neutron in their nucleus—now enable scientists to date reliably much more recent biological items such as wood or cloth or animal remains that range in age from a few hundred to tens of thousands of years.

But let us return to the Earth itself. Its age is about one third of the probable age of the universe, which we discussed in Chapter 4 as being about 15,000 million years from astronomical data. This relationship seems very reasonable and allows for the sun an age of perhaps 10,000 million years, give or take a few thousand million years. The aim of geological science is to discover all that has happened to the earth in that immense time, to show how its rocks and mineral deposits were formed, and to determine what shaped its continents and oceans. Geology also throws light on the origin and development of life. Let us take these topics in turn.

The Structure of the Earth

When the Earth first collected itself together out of the dust and gas left over from formation of the sun it was very hot. The main reason for this was that all the material crashed onto the growing surface at immense speed, turning its kinetic energy, derived from gravitational potential energy, into heat, but there was also heat generated by the slow decay of radioactive material. Since heat was lost from the surface as infrared heat radiation, the interior of the Earth was a good deal hotter than the surface, and nearly all was molten. This allowed materials that were heavier to sink towards the centre of the Earth to form a core that is composed largely of iron, while lighter materials floated. Over the immense times involved, the structure of the Earth settled into four zones: the inner core which seems to be solid, the outer core which is liquid, the solid but slightly plastic mantle which contains most of the Earth's material, and a very thin solid crust. Even the deepest mines and drill holes have not yet penetrated through the crust to the mantle, so that most of our information about the deep structure of the Earth comes indirectly from a study of the way in which earthquake waves propagate and from a knowledge of the way in which rock materials behave under conditions of very high temperature and pressure—the temperature near the centre of the Earth is still about 4000°C. Lying above the crust are the oceans and the atmosphere, which consist of the lightest and lowest-melting materials formed from the original mixture.

Most of our practical interest centres, of course, on the regions very close to the surface of the Earth—the continents, the oceans and the atmosphere—for those are the only parts yet accessible to us. Let us defer the oceans until later in this chapter, and the atmosphere to the next chapter, and look at how the continents formed.

It is fairly realistic, though not very flattering, to compare the Earth's crust with the scum that forms on the top of hot cooking pots, which clumps together into big islands that slowly move around on the surface of the liquid in the pot. The islands of crust, which we call continental plates, float on top of the heavier material of the mantle and move around extremely slowly. This is not a contradiction of our earlier statement that the mantle is solid—it flows extremely slowly under the stirring action (convection) of the heat in the core, much as the solid ice of a glacier flows slowly down a mountainside under the force of gravity. The analogy to a cooking pot is therefore quite a good one, since the movement in the pot is driven by the heat of the stove on which it sits.

We can, in fact, now measure the speeds with which the continental plates

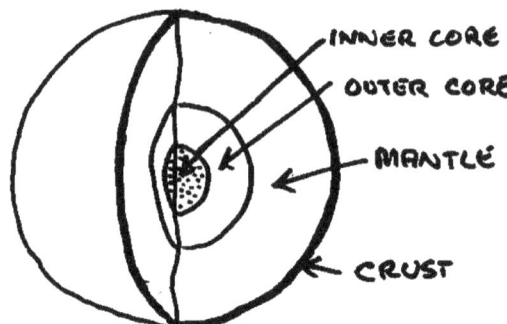

The Earth has a solid inner core, a molten outer core, a solid but plastic mantle, and a thin solid crust floating on the outside. Above the crust on which we live are the oceans and the atmosphere.

The Earth's continents are massive floating plates which drift slowly across its surface. Five hundred million years ago, many of the continents were clustered together to form a giant supercontinent called Gondwana, which later broke up.

are moving, using very precise measurements from satellites, and we find typical speeds of about a centimetre per year, which means only about ten kilometres in a million years. The Earth's crust has been solid for several thousand million years, however, which means that the continents have been able to drift thousands of kilometres. A great deal of evidence from matching rock types, as well as evidence from their shapes, indicates that about 500 million years ago Antarctica, Australia, South America and Southern Africa were all joined together into a big continental mass, now given the name Gondwana, which later broke up and separated. Other plates have moved together and collided, and it seems clear that India, for example, is a smaller plate that has collided with the larger plate we might call Asia, the slow and steady impact of the collision raising the steep mountains of the Himalayas and the high plateau of Tibet.

Landforms

As the continental plates move apart, they leave the deeper basins in which water has collected to form the oceans. Along the floors of the oceans there are tear-lines through which flows molten material to fill up the gaps. Such huge tear lines or rifts have been identified and mapped along the floors of the Atlantic and Pacific Oceans. Other lines along which the crust is under great stress occur, for example, around the rim of the Pacific Ocean and give regions in which earthquakes and volcanoes are common—Japan, New Zealand, the Andes and the Rocky Mountains. These have played a large part in developing the topography of such areas.

The origin of earthquakes is fairly clear—with all this action going on, it would be surprising if the crust did not suddenly crack and shift here and there, releasing strain energy that may have been built up for centuries. On the scale of the Earth, these are very minor events, but if we happen to have built our cities, close to the fault line that slips in the earthquake, then the devastation can be immense. Australia is fortunate in not having major active faults running through it, as have Japan and California, but even here there are occasional serious earthquakes that have caused considerable damage to cities such as Newcastle.

Volcanoes and flows of molten rock have played a great part in building up the landscape too. Some types of rock have a rather low melting point—a few hundred degrees Celsius—and exist as large reservoirs of molten material deep in the crust. When the crust is weakened because of plate motion, it is relatively easy for some of this material to be squeezed out. What happens then depends

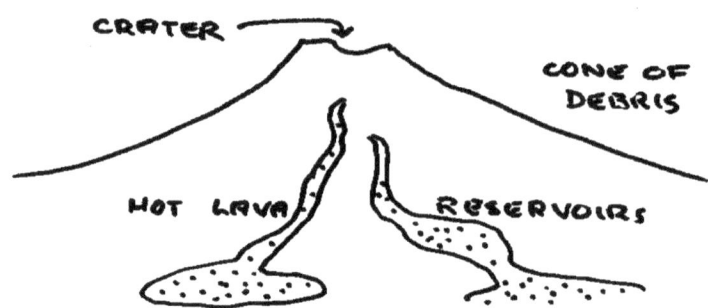

Volcanoes are mountains formed by the outflow of molten lava, which is just a rock with rather low melting temperature, from reservoirs deep in the Earth's crust. Some volcanoes eject gas and ash rather than lava.

upon exactly what sort of molten rock it is. Some is very viscous and builds up high mountains with a crater on top that occasionally gets blown off when the internal pressure becomes too high. Some, however, is quite fluid and runs as uniform sheets over huge areas of the land surface. The traditional shapes of volcanoes are familiar from photographs, one of the most famous and beautiful being Mount Fuji in Japan. More common in Eastern Australia are the basalt lava flows that are often exposed near the headwaters of coastal rivers. This can often be seen in separate flows up to tens of metres deep, and usually each flow has split into roughly hexagonal columns as it has cooled, rather like the patterns in drying mud in a lagoon.

Along with all this building activity in the landscape, there has also been continuous destruction by the forces of erosion. The mechanical forces exerted by freezing ice, running water and fierce winds, added to the occasional destructive violence of eruptions and earthquakes, continually break down the landscape and flatten it, with the debris accumulating in shallow basins, often under stretches of water. Rock materials are also attacked by water and air and decomposed to soil particles which are themselves eroded by wind and rain and washed down into the oceans.

With all this action going on, it is not surprising that the continental plates have tipped, buckled, risen and fallen over the course of their history as well as just drifting. Some of the distortion has been quite local and can be seen

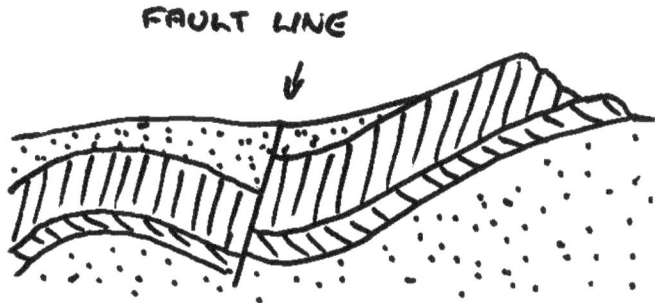

Bands of tilted sediments are often visible in road cuttings. Sometimes the bands are so thick that they can be seen in the shapes of hills.

in road cuttings, some can be seen in the great tilted stone blocks of mountain ranges, and some leads to great flat basins filled with the accumulated debris from ground-down mountains. It is a fascinating occupation to look at the landscape through which you pass and to try to fit its recent geological history into this general picture.

Rocks and Fossils

Geologists studying all this in detail have classified the materials from which the Earth is built in many ways. Here we discuss only the materials of the accessible crust. Rock is the solid building material from which the crust is built, and there are two main classes based upon origin. Primitive rocks are called igneous, reflecting their origin as molten material deep in the Earth, although some of them may be of recent formation in volcanoes. Such rocks are generally hard and, if we look at them closely, made up of small crystals all rigidly cemented together. An excellent example in which the crystals are easily visible is the grey or pink granite used for facing many city buildings, while another in which the crystals are much less obvious is the basalt that is crushed up to make road surfaces.

The second type of rock is called sedimentary, and is a secondary rock made of the compressed remains of other rocks that have been eroded away into small grains or pebbles by wind or water and has the settled to the bottom of a lake or sea. A typical example is the sandstone rock so common around Sydney, visible as huge cliffs in the Blue Mountains and used as a building material in Colonial buildings in many State capitals. Some sedimentary rocks are much coarser than this and have fine pebbles in them, while some are much finer and appear to be hardened mud. Another important sort of sedimentary rock is that made from the shells or skeletons of tiny marine creatures, settling for millions of years to the bottom of quiet lakes or seas to form either limestone or marble. Most of these sedimentary rocks are rather soft compared with igneous rocks, but some of them may have been violently heated by volcanic action long after their formation to partly fuse them into hard "metamorphic" rocks, the word meaning "changed in form".

When sedimentary rocks form in lake beds, the skeletons of animals living in the lake become embedded in the sediment at the bottom of the lake and preserved from decay. Sharp floods may even bring down tree leaves and small branches that become similarly preserved. When the lake dries up and the sediments are pressed hard into rocks, these animal or plant remains are preserved as fossils and can be revealed by splitting open the rock along horizontal planes. Much of our knowledge of the early development of life comes from such fossils—from early plants and tiny sea creatures to the preserved skeletons of huge dinosaurs. If the rock beds in which the fossils have been preserved have not suffered too greatly from volcanic action or other disturbances, then the sequence of life forms can be built up, with the earliest at the bottom of the sediments. By comparing such fragmentary records from many parts of the world, a fairly complete history of the development of life on this planet can be constructed, as we discuss in a later chapter.

Minerals

A rock is a complex material often consisting of many visibly different types of crystals. The number of different materials found in the crystal grains is, however, limited, and each is a pure substance that is called a mineral. Examples of minerals are silica which is ordinary sand, mica which occurs as tiny stacks of sheets, all the gem stones such a diamond, ruby and garnet, and less commonly seen materials such as iron ore, zinc ore, aluminium ore, and even metallic gold. What we regard as the important minerals depends upon our point of view—to a geologist they are the major minerals that make up the rocks of the Earth, while to an economist they are the minerals from which valuable materials can be recovered.

The mining and processing of minerals is one of the oldest activities of civilised humans, though not as old as the growing of crops or the making of stone and ceramic utensils. While gold, found "native" as a metal, was used for rich ornaments, the earliest useful minerals were the ores of copper and tin. These were easily smelted in wood fires to give the pure metals, which could then be melted together in roughly equal parts to give the much stronger bronze. Iron was a later development because of the much higher temperature and more complex procedure needed to recover it from its ore. Today we can recover any metal we want to from its ores, and materials such as aluminium and titanium are becoming increasingly important. Aluminium is an interesting material because it is immensely abundant in the Earth's crust, though not usually in a convenient form, and is very useful because of its lightness and resistance to corrosion. Modern technology could not function without the use of a wide range of pure materials derived from minerals in the Earth's crust. For simple structural uses many alternatives exist, but for sophisticated technology there is often only one suitable material.

Fossil Fuels

Among the most important things for modern industrial society that are available form the Earth's crust are the fossil fuels—coal, oil and natural gas. All these were derived from the prolific plant life that covered the Earth's surface about 300 million years ago and have been preserved under beds of sedimentary rocks. The conditions for preservation were not found everywhere, and much of the vegetation simply decayed. When it was preserved, the geological conditions and the nature of the original plant material determined whether it was as coal or oil. Coal occurs as solid material in seams up to five metres thick, while oil and gas are trapped in porous sedimentary rocks. The brown coal in Victoria's LaTrobe Valley is very different from the black coal of New South Wales and Queensland, and the light oil from Bass Strait wells is very different from the thick black oil of the Middle East. Natural gas, which is really the lightest part of the oil with a boiling point below room temperature, is generally produced along with the heavier oil, but is retained only if the geology of the region provides a reservoir of porous rock with a dense domed roof through which it cannot escape.

Although coal, oil and natural gas were originally derived from the energy of sunlight through the agency of plants, they are clearly not renewable resources—

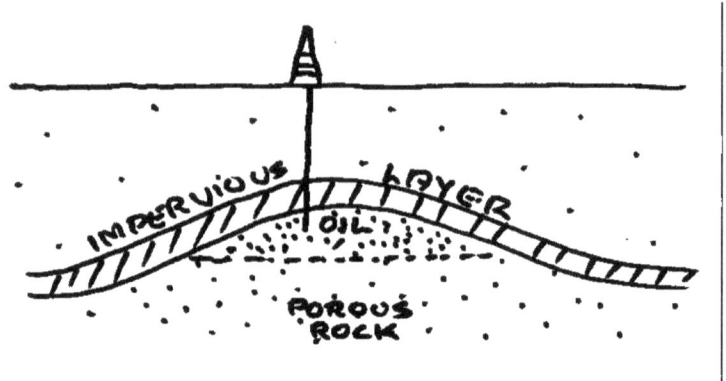

Oil and natural gas are the remains of old vegetation. They are trapped in porous rocks covered by impervious layers that prevent their escape. Wells drilled through the impervious layer allow us to recover the gas or oil.

once we have used up the Earth's present stock, there will be no more. Both coal and oil are the sources of extremely important chemicals, as well as being the world's major sources of energy, and it is almost criminal that we should be squandering them as we do at present. This is a problem to which we return later.

The Oceans and Ice Caps

Water is an abundant material on the surface of the Earth, though very unevenly distributed, and is clearly essential for the maintenance of all forms of life. The fact that it exists in solid, liquid and vapour phases and changes between these very extensively under the conditions on the Earth's surface is also responsible for many of the geological processes and much of the biological diversity that we find.

Underground water is a very important commodity in many parts of the world, including much of Australia. In the form of artesian water, it is a huge and slowly-flowing reservoir trapped in porous limestone rocks between impervious strata. Australia has several artesian basins, the largest of which underlies all of Western Queensland and the northern parts of South Australia and New South Wales. There are other important artesian basins in the Northern Territory,

Western Australia and Western Victoria. The water supply in the artesian basins is not a static thing, like oil reserves, but is replenished by rainfall and river soakage in the wetter parts of its area. Such groundwater is essential for life in many parts of outback Australia, and farmers rely heavily upon bores drilled to bring it to the surface. Unfortunately the number of bores drilled has increased so greatly that the supply is being used faster than it can be replenished, and the total flow of artesian water in Australia is now only half what it was in 1920, though the number of bores has doubled.

Groundwater can also cause important problems when it lies close to the surface, rather than being confined in artesian basins. As the water from rainfall or irrigation moves through the soil, it dissolves out salt. This is no real problem so long as it remains well below the surface, but the clearing of trees from the land or inefficient irrigation practices can raise the water table right to the surface in low-lying places. This has two effects. In the first place, it can kill off any remaining trees that are not adapted to having their roots under water. More seriously, the water brought to the surface evaporates and deposits its load of salt to make salt pans in which nothing will grow. These spread and kill more trees, making the problem worse and worse. Such salination of once good land is a major management problem in dry regions of Australia.

The oceans cover rather more than half of the Earth's surface, and the deepest part of the ocean, the Mariana Trench in the Western Pacific, is 11,000 metres deep, which is more than the 8,848 metre height of Mount Everest above sea level. There is not a great deal of mixing between deep-ocean water and surface water, but the oceans have circulation patterns that are important to controlling world weather and climate.

One of the major physical features of the oceans is the tides, which occur twice each day at times that shift with the phase of the moon. The tides are, in fact, due to the gravitational influence of the moon on the Earth, with one high-tide bulge pointing a little behind the position of the moon and the other balancing it on the opposite side of the Earth. The sun also creates tides on a rather smaller scale, but when the two line up at the time of new moon or full moon the "spring tides" that they produce are particularly large. Because so many of the world's major cities are built on the coast, and are therefore only a little above high-tide level, the stability of the average sea level is important. Concern about possible rises in sea level is a part of general concern about climate change and the "greenhouse effect" to which we return in a later chapter.

Finally, when discussing the water resources of the planet, we must not forget the ice caps of the Arctic and Antarctic. Between them they hold about 5 percent of all the water on Earth, and more than 90 percent of the Earth's supply of fresh water. The Arctic ice, though extensive, floats on the sea and is not very thick. It can almost be neglected in relation to the huge ice mass of the Antarctic, which lies as a cap up to 5000 metres thick over all the interior of the Antarctic continent, which is itself bigger than Australia. This huge cold white mass extends to nearly twice its normal area in the winter because of growing sea ice. Although this does not affect the stored ice volume much, it has a very important influence on the Earth's weather and climate and on the circulation of the oceans. Australia maintains scientific stations on the Antarctic continent itself, as well as on Macquarie Island, to study physical processes in the region and the biology of Antarctic wildlife.

7

Weather and Climate

"With hey, ho, the wind and the rain." Shakespeare "Twelfth Night"

Three of the most important things to life are sunlight, air to breathe and water to drink. With these three, plant life can prosper and provide food for other life forms to eat. The atmospheric sciences are devoted to the study of the atmosphere generally, but particularly of weather and climate, and can be collectively called Meteorology, though this term is often used in the more restricted sense of weather prediction.

Because the atmosphere is constantly in motion and never repeats its patterns, much of the effort in atmospheric science must go into collecting data on important measurable quantities such as temperature, atmospheric pressure, wind direction, cloud cover, humidity and rainfall. To make matters even more complicated, these can all vary differently with height at different places, so that balloons carrying measuring instruments (radiosondes) have to be released every few hours to make measurements all the way up to high altitudes and radio them back to earth. To obtain enough information to predict changes in weather accurately, this information must be collected continuously over as much of the Earth's surface as possible and supplied immediately to meteorologists making the forecasts. Modern radio communication has made this possible, and there is world cooperation in collecting and processing the measurements. The Australian Bureau of Meteorology in Melbourne is one of three World Meteorological Centres, the other two being in Europe and America, and is charged with collating the data and making forecasts for the whole southern hemisphere.

Of course, weather satellites have made an immense contribution to the whole enterprise over the past thirty years. These satellites are in orbits above the Earth's equator at such a height that they complete an orbit once every 24 hours, and so remain exactly over the same place on the Earth's surface—this is called a geostationary orbit. Each satellite is able to see nearly half of the Earth's surface, though the view is much better near the centre of this area than near the edges. There are many such satellites covering different parts of the globe, the one that covers the Australian region being a Japanese weather satellite. Each satellite gives an excellent picture of cloud cover, such as we see on television or in the newspapers, and the more modern ones are able to make other measurements as well.

Driving the Weather

The first question to ask about the weather and climate of Earth is what determines the average temperature. The answer is quite plainly the radiation we receive from the sun, since measurements show that Mercury and Venus, which are closer to the sun than is Earth, are both a good deal hotter, while the planets further from the sun are all progressively colder. Details of the balance are, however, more complex than this. The energy of the sun's radiation, at a temperature of about 6000°C is predominantly in the visible range to which the atmosphere is fairly transparent. Some of this radiation is reflected from clouds and from the surface of the Earth, but a good deal is absorbed by the ground, by plants, and in the oceans, warming them up. At the same time, all the surface of Earth is radiating heat itself with energy concentrated in the far infrared because of its low temperature. These two energy flows have to balance when the earth is at a steady average temperature, allowing for fluctuations between day and night and between winter and summer.

Now while the atmosphere is transparent to visible radiation, it is far from transparent to infrared radiation, and a good deal of that radiation emitted from the surface is absorbed in minor gases such as water vapour, carbon dioxide and methane that are present in small quantities in the air. This is a good thing from our point of view, for if it were not so then the average temperature of the Earth would be colder by perhaps 20°C and we would all be freezing! This is called the "Greenhouse Effect" because something similar happens in a greenhouse—the glass window panes, as well as letting in the sun and keeping out the wind, also keep in the infrared radiation and allow the plants to keep warm even in the winter.

With the average temperature settled, we now turn to weather. In the final analysis, the sun is also responsible for all the Earth's weather. It is always more

Weather and Climate

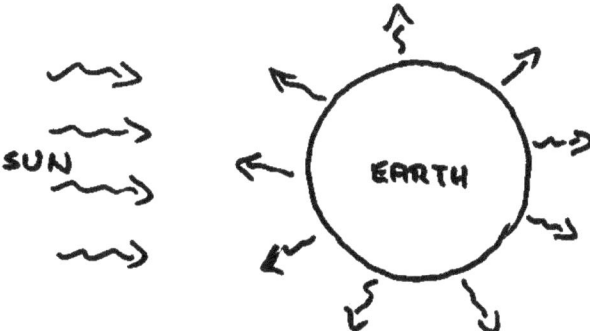

The average temperature of the Earth is determined by the balance between incoming radiation from the sun and infra-red (heat) radiation emitted by the warm surface of the Earth. More radiation is absorbed than emitted near the equator because the sun is overhead, but near the poles the amount absorbed is less than that emitted.

or less overhead at midday in the tropics near the equator, so that this region receives a large amount of heat energy which warms both the Earth's surface and the air above it, as well as evaporating water from the land and the sea into the air. The warm air rises, because it is lighter, and spreads out at high altitude away from the equator. This air is itself replaced by cooler surface air flowing towards the equator, and the two flows join up south of the tropics, with the upper air coming down to the surface again, to form an immense vertical circulation cell. The whole thing is not an even flow, however, but more like the flow patterns that you can observe in a saucepan of soup boiling on the stove.

This circulation pattern has an important influence on rainfall. As the hot air rises in the tropics, it cools and water vapour condenses out to form clouds which produce heavy rain, largely as tropical thunderstorms. At the other end of the pattern further from the equator, the now-dry upper-level air warms as it descends and reaches the surface as hot dry winds, producing a ring of deserts, or at least very dry areas, around the whole world. In Australia, Darwin and Cape York are in the wet tropic region of high rainfall, while much of the centre of the continent is in the dry belt of latitudes. The whole pattern shifts with the seasons because of the tilt of the Earth's axis, so that the belt of low pressure and the high rainfall moves south in the southern summer every year, bringing

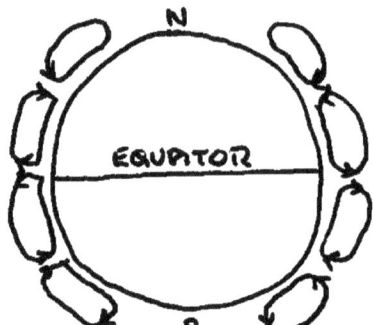

Hot moist air near the equator rises and moves towards the poles in the upper atmosphere, to be replaced by cooler drier air moving towards the equator near the surface. This forms a large circulation pattern which has a major influence on the weather. Other similar cells exist closer to the poles.

monsoon rains to the northern and central part of Australia, with drier weather farther south. In the southern winter, the monsoon belt moves into the northern hemisphere.

Farther still away from the equator, the vertical circulation pattern repeats at higher latitudes, giving another low-pressure belt where the air rises at latitudes rather to the south of Australia. The rainfall is not so high here as in equatorial regions, however, because of the lower temperatures. These patterns of air motion can be seen on weather maps as patterns of high and low pressure—low pressures correspond to rising air and high pressures to descending air. Normal air pressure is 1013 hectopascals (hPa)—the same unit used to be called a millibar—so that you can judge the deviation in pressure from that, but in addition the high and low pressure regions are always marked "High" and "Low" on the map.

The pattern of air movement is, however, much more complicated than this because of the rotation of the Earth. This causes the air to be deflected as it moves north or south so that, in the southern hemisphere, the air moves in a clockwise direction as it flows into a low-pressure region and in an anticlockwise direction as it flows out from a high-pressure region. The directions of circulation are opposite to this in the northern hemisphere. The strengths of winds

Weather and Climate

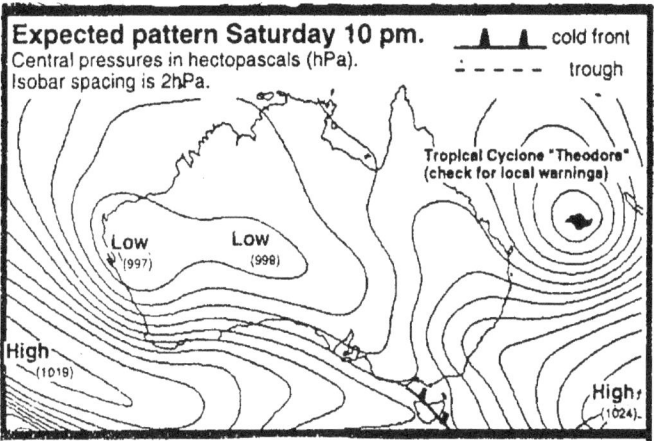

A typical summer weather map for Australia. Note the regions of high and low pressure, the tropical cyclone off the Queensland coast, and the cold front south of the continent. Patterns such as this move slowly from west to east across the continent but change in detail as they move.

round a low-pressure region depend upon how rapidly the air is rising, which is in turn indicated by how much the pressure is below the standard value of 1013 hPa. This is indicated on a weather map by the numbers giving pressure, but also shows up as the contour lines—isobars—that are drawn through places of equal pressure. The winds blow along the isobars in a clockwise direction, for a low-pressure region in the southern hemisphere, and their strength is proportional to the closeness of the isobars—a map with a lot of isobars close together means very windy weather. Extreme cases of this occur in tropical regions and lead to violent and widespread storms called tropical cyclones in the southern hemisphere and typhoons in the north.

Apart from these cyclones, which move rather erratically, the weather patterns progress in a fairly orderly manner from west to east in both hemispheres. It takes about a week for the weather in Perth to travel across to Sydney, though this can vary considerably. It is the job of meteorological forecasters to predict the rate of movement of the pattern, whether the lows and highs strengthen or weaken as they move, and what this means in terms of temperatures, winds and rainfall.

An important feature of the weather in temperate rather than tropical regions is weather fronts—usually cold fronts in Australia. These are places where a new air mass from the south is flowing into a low-pressure region, circling as it progresses. Cold fronts are usually associated with rain and clouds, and of course with cooler air temperatures. They are marked on weather maps as a curved line with triangular protruberances. Warm fronts are rarely seen in Australia, but are a common feature of weather in the northern hemisphere.

Clouds and Rain

Clouds are an important feature of the Earth's weather, and it is interesting to look at them in a little detail. We know that they are water that has condensed as moist air rises and cools, but just what are they like, and why do some of them rain and some not? We don't yet know all the answers, but we can go a good way towards answering these questions.

The shapes of clouds depend largely upon the way in which they were formed. If the moist air is lifted gradually and over a large area by having cold air flow in underneath it or by flowing up a long gentle mountain slope, then condensation occurs smoothly and we get sheets of cloud with fairly flat tops and bottoms that are called stratus clouds (stratus meaning something flat). Their thickness may range from just a few tens of metres to hundreds of metres. On the other hand,

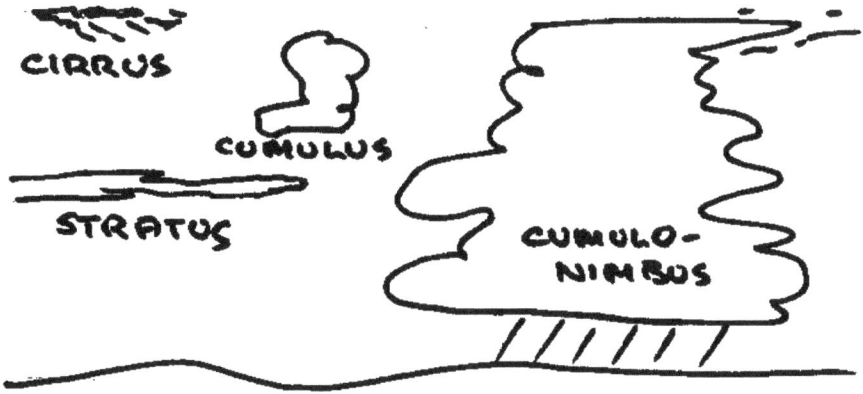

Clouds are formed when warm moist air rises and its moisture condenses to tiny water droplets. There are many types of clouds depending upon the temperature, humidity and wind at a particular time.

if the air is simply heated by the hot ground, then it does not rise uniformly, but starts to do so in isolated places which then act rather as funnels for surrounding air. The clouds that form in this case are isolated packed-up structures in which the air is going up rapidly in the middle and down slowly around the edges. They are called cumulus clouds (cumulus meaning a heap), and may have thicknesses from base to top ranging from hundreds of metres up to several kilometres. Of course we may have all situations between these extremes and cloud types such as stratocumulus. Two particular cases are worthy of mention. The first is that of cirrus clouds (cirrus meaning a tendril or hair), which are very high lacy clouds; at the other extreme we find towering black thunder clouds with an anvil or halo of ice crystals blowing off the top, which are called cumulonimbus (nimbus meaning a halo).

Clouds are formed from water droplets that are too small to fall through the air at any speed. Typically these droplets are about 10 micrometres, or one hundredth of a millimetre, in diameter and have fall speeds of only a few centimetres per second. Their concentration is typically a few hundred per cubic centimetre. Many processes contribute to collecting these tiny droplets into raindrops, which typically have diameters of a millimetre or more and so represent the combination of thousands of cloud droplets. The cloud droplets may simply collide and coalesce, but this is an inefficient process for small droplets and works better if their initial sizes are rather larger. Alternatively, if some of the droplets freeze in the cold top of the cloud, then these tiny ice crystals can grow at the expense of surrounding water droplets, melting later to raindrops as they fall into warmer air. The more water there is in a cloud, the larger its droplets, and the greater its height from base to top, the more likely it is to produce rain. Even a violent rainstorm, however, depletes the water content in a cloud only a little. Most of the droplets simply evaporate again when the air containing the cloud becomes warmer as it sinks later on in its history.

Climate

While weather is something that varies from day to day and from month to month, climate is a much longer-term average. In Australia, even successions of drought years are regarded as variations in the weather rather than in the climate, and the climate is described by the fraction of years in which we experience drought in a given place. On the other hand, the fact that we have drought years and flood years, rather than a whole sequence of average years, is part of the climatic pattern of Australia which makes it different from the more equitable climate pattern of Europe.

Many things go to make up climate and its variability. One is persistence—if you predict tomorrow's weather will be pretty much like today's and that next year will be pretty much the same as this year, then you are likely to be right more than half of the time. There are sudden changes, however, some of which are understood and some of which appear to derive from random events that are characteristic of weather patterns. One strong influence on Australian weather and climate on a scale of years is the temperature distribution in the Pacific Ocean—the El Niño phenomenon. When the ocean temperature is higher near Australia than near South America, then the average air pressure is lower near Australia than near South America and we tend to have wet seasons here. The state of the ocean typically persists for several years before making a fairly sudden change to the opposite state, and our weather does likewise. The circulation of the world's oceans is very closely connected with climate and weather.

Climate has changed very significantly over long periods, as we can tell from geological studies. At times there have been forests in Antarctica, where all is now covered in ice; there have been glaciers in Southern Australia and the highlands of New Guinea, and much of Northern Europe has been covered by ice. Indeed from geological evidence, and particularly from examining cores drilled through the deep ice of Antarctica which can yield very detailed temperature records on an annual basis, we know that successions of ice ages and warmer climate alternate with separations of order 100,000 years. We are now in one of the interglacial periods in which the temperature is relatively high, and the indicators in the ice-core record are that it will go higher. Mind you, the variation in average temperature is not extreme—a bare ten degrees difference in average temperature separates the depths of an ice age from the warmth at the peak of an interglacial period!

Climate Change

Because weather is so important to agriculture, forestry and food production generally, it is vitally important to know whether it is likely to change so that we can plan for this. Even more important is to know whether human activities are influencing the climate, or likely to influence it in the moderately near future, so that we can perhaps do something about this.

The immediate thing we note when we think about it is that human activities are releasing a great deal of carbon dioxide into the air by burning fossil fuels such as coal and oil, and that the rate at which we are doing this is increasing. Actual measurements of carbon dioxide concentration in the atmosphere far away from cities show that this concentration has increased by something like 50 percent over the past hundred years and suggest that it will have doubled by the year 2020. The concentration of methane is also increasing because of agricultural activity and the release of gas from oil wells. Now these are two of the most important gases responsible for the greenhouse effect in the atmosphere, so what is their increase likely to do to average global temperatures?

The answer is not simple, because the weather system of the Earth is very complex. Certainly more carbon dioxide means an enhanced greenhouse absorption, which should raise the temperature significantly. But that in turn will increase evaporation from the oceans, which will increase the amount of cloud, which will reflect more of the sun's radiation, which will tend to cool things! The answer depends upon a delicate balance. Large and carefully constructed computer models taking all these effects into account suggest that the temperature will probably rise by between one and three degrees if the carbon dioxide concentration is doubled, but that is simply a best estimate based on present evidence and is not conclusive.

Similar uncertainty is attached to the question of what will happen to the sea level. Many major cities are built just above the high-tide line and any significant rise could be disastrous. Again, if the temperature rises then the top layers of the oceans will warm a little and expand, which could raise the sea level several metres. At the same time there will be increased snow fall in Antarctica which may more than balance the slight melting around its edges and have the effect of lowering the sea level. Again, a best estimate using all presently available data suggests a rise in sea level of about 30 centimetres in the next 50 years, but the conclusion is still tentative.

Finally, what will happen to the rainfall? This may well be the most important question of all. Rainfall predictions are much more difficult than either temperature or sea-level predictions and the answer is far more complicated because rainfall is likely to increase in some places and decrease in others. Different models have a reasonable measure of agreement, but the details are not reliable, Australia is probably big enough that a decrease in one place will be balanced by an increase in some other area, but in heavily populated tropical countries the consequences of change could be disastrous.

So there is no definite prediction yet, but the danger flags are out, and we disregard them at our peril, or rather at the peril of our children and grandchildren. The best advice we have comes from scientific models—other than that there is nothing—and they all suggest that we should reduce our carbon dioxide emissions on a global scale as rapidly and as comprehensively as possible. What is more, of course, this advice makes sense on resource grounds even if there were no greenhouse effect to worry about. We shall look at the sort of trade-offs that such a decision involves in a later chapter.

Atmospheric Pollution

Carbon dioxide and methane are, however, not the only gases we have to worry about in the atmosphere—indeed apart from their likely effects on climate they are relatively harmless. Much more important are some of the more corrosive and toxic materials released into the atmosphere by human activity. Many of these pollutants are produced by industry and many by transport.

Some of the most unpleasant pollutants are the acidic gases such as the oxides of nitrogen that are produced when fuels are burnt at high temperatures, as in automobile engines and in many industrial processes. These gases make the air of cities irritating to our eyes and throats and have similar effects on plants and buildings. Cathedrals and other great buildings that have stood for centuries are now crumbling as the acid air eats at their stonework, and the same thing happens to the metal roofs of more modern buildings. Rain washes the acid gases out of the air, but this acid rain is beginning to have a devastating effect on trees throughout the world. Exhaust gases from motor vehicles also react in the sunlight to produce a haze of photochemical smog over our cities which makes them much less pleasant places in which to live.

The effects of some pollutants is less obvious but just as important. The sun's light contains a considerable amount of ultraviolet radiation that is filtered out by its interaction with ozone (O_3) generated from ordinary oxygen (O_2) high in the atmosphere. Many pollutants can interfere with this process, but some of the worst are the refrigeration chemicals called freons or CFCs (chlorinated fluorocarbons) which can be released from leaking or discarded refrigerators. These gases break up the ozone molecules but are not themselves destroyed in the process, so that they typically remain in the atmosphere for ten years or more. The effect of these gases on ozone concentration is becoming quite marked, particularly in high southern latitudes where an "ozone hole" that grows in extent each year has been found. Less ozone above our heads means that more ultraviolet radiation reaches the surface of the earth, and the cells or our bodies have not evolved to cope with this. The UV radiation damages skin cells in subtle ways, and this damage can lead to the growth of dangerous skin cancers.

Everywhere around the world steps are being taken to reduce the release of all these materials that are polluting our common atmosphere, but the task is not simple. Primitive or worn-out equipment produces more pollution than well-designed new equipment, but replacement costs are high; new refrigerants are being developed to replace CFCs, but they are more expensive; nuclear power could replace coal in electricity generating stations, but it brings problems of safety and waste disposal. At least politicians and economists are becoming aware of the threats that face us all and beginning to take some action.

8

Magnetism and Electricity

"Electricity is of two kinds, positive and negative. The difference is, I presume, that one comes a little more expensive, but is more durable; the other is a cheaper thing, but the moths get into it."
Stephen Leacock "A Manual of Education"

Magnetism and electricity are so important to everything we do in our modern technological society that they merit a chapter to themselves. This is even more necessary since they are both rather mysterious things that cannot be seen or felt—unless we are unlucky enough to get an electric shock. Despite, or perhaps because of, their mysterious nature, magnetic and electrical phenomena were among those first studied by scientists of the ancient world. The Greek philosopher-scientist Thales of Miletus (640–546BC) was familiar with the fact that amber, when rubbed, will attract light objects such as pieces of paper, and that a certain mineral, now called magnetite, had the power of attracting iron. The Chinese, probably well before this time, were similarly familiar with magnetite and also knew that a piece of magnetite suspended from a thread always pointed in a fixed direction. They used this primitive magnetic compass when travelling.

The mysterious nature of magnetic and electric phenomena has also led to a great deal of superstition, some of which persists in a mild form today. Bracelets and other charms of amber or copper and small magnetic stones are believed by some to have healthful properties, though the same people are often fearful of the supposed danger of electric and magnetic fields from power lines. There does not seem to be any evidence that these charms have any effect on the human body. We cannot be so sure about the fields near power lines and electric machines, however, since these can be quite intense, but most of the studies of people living near power lines are inconclusive because so many other factors related to occupations and living habits also enter. Certainly, however, even very strong magnetic fields seem to have no harmful effects for short times. Electric fields are similarly harmless—even the fine-weather electric field at the Earth's surface is about 300 volts per metre—but of course electricity does present real dangers in other forms, as we discuss later.

Magnetism

Magnetism is a property of the atoms of certain metals, particularly iron, cobalt and nickel but also others including the "rare earth" metals such as samarium and neodymium. This does not mean that all magnets have to be metallic, but it does mean that they must contain atoms of these particular materials. A magnetic atom is like a small arrow in that it has a magnetic "head" and "tail" which define the direction in which its elementary magnet points. The important thing about magnets is that the head end of one magnet attracts the tail end of another, so that they like to line up. Conversely, two head ends will repel one another. We can demonstrate this easily with two bar magnets.

If we take an ordinary piece of iron, the atomic magnets are aligned into little clumps, but the clumps themselves point in random directions so that there is no head or tail end to the piece of metal as a whole. If, however, we bring a strong magnet near to a piece of iron, it will change the directions of some of the magnets and leave the iron magnetised to some extent. This is how the mineral magnetite became magnetised. The Earth, as we see later, has a weak magnetic field from its iron core, and this magnetised the magnetite mineral, which is a sort of iron oxide, while it was still hot. As it cooled down over the years, it retained this magnetisation to be a "permanent" magnet. Permanent magnets are made similarly today, using strong magnetic fields to align the atomic magnets and then setting them in place as the material is cooled.

We mentioned that the Earth's core generates a weak magnetic field and behaves therefore rather like a large magnet. The exact mechanism by which the field is generated is not completely clear, but it involves the fact that the outer core of the Earth consists of molten iron, rotating with the Earth. The axis of the magnetic field does not line up exactly with the rotation axis of the Earth, however, and the present position of the north magnetic pole is in the north part of Canada, while the south magnetic pole is opposite it in Antarctica. In Australia the direction of "magnetic north" which we find from a compass is about 15° to the east of "true" or geographic north. The magnetic

A magnet is surrounded by a magnetic field, the strength and direction of which can be made visible by sprinkling iron filings on a piece of paper with a bar magnet underneath it. The Earth's magnetic field has a similar form.

poles do not stay precisely fixed, however, and this "magnetic deviation" from true north changes slightly over periods of tens of years. Indeed, geological evidence from magnetised rocks indicates that the Earth's magnetic field has even reversed rather suddenly from time to time, the most recent occasion being about 100,000 years ago.

Since one end of a magnet will point to the north magnetic pole of the Earth, we call this a "north-seeking pole" or simply a north pole. This is perhaps a little confusing, because it means that the north magnetic pole of the Earth is actually a "south pole"! Let us not worry about that, however.

A magnet is surrounded by a "magnetic field" which gets weaker as we get farther away. We can make the strength and direction of this field visible by placing the magnet under a sheet of white paper, sprinkling iron filings on the paper, and then gently tapping the paper. The filings are magnetised by the larger magnet under the paper and line up head-to-tail along the direction of the magnetic field. The Earth's magnetic field looks just like this, with the magnet buried at the centre of the Earth.

While we are not aware of the Earth's magnetic field unless we have a magnetic compass, it does have some other effects. One of the most spectacular is the aurora australis which can sometimes be seen in southern parts of Australia. (The same phenomenon in the Northern hemisphere is called the aurora borealis.) What happens is that charged particles from the sun are trapped in the Earth's magnetic field high above the atmosphere and funnelled down towards the magnetic poles. Here they interact with the upper levels of the atmosphere and produce a glow rather like the glow of a neon advertising sign, though not so bright. The southern hemisphere aurora is called the Aurora Australis while that in the northern hemisphere is the Aurora Borealis.

Electricity

As the ancients found when they rubbed amber, electricity can be produced by friction. The reason why this is possible is that, as we see in the next chapter, most of the particles that make up the atoms of matter have electric charges, and the friction simply rubs off some of them onto the cloth, leaving the cloth with electricity of one sign and the amber with the opposite sign. The fact that there are two different signs for electric charge and that they can be separated makes electricity very different from magnetism—if we break a magnet into two pieces then we get two smaller magnets, each with a north and a south pole, and there is no such thing as an isolated magnetic pole.

Although we may not be used to going round rubbing pieces of amber, we have all experienced frictional electrification when we get a mild electric shock from sliding quickly across the seat of a car and touching the door handle, or from doing something similar on a new carpet in dry weather. We may even see the tiny spark as the separated electric charges jump back together again. The reason that the shock makes us jump is that our nerves all operate by sending tiny electric signals along their fibres, so that an electric shock automatically tells our muscles to contract.

A word about electric shocks is appropriate here. The important thing is the amount of electric charge that flows through our body, and that depends on the voltage that is applied. We look at just what voltage means presently, but we are all familiar with the numbers—a dry cell battery for a torch or a transistor radio is either 1.5 or 9 volts, while the household electricity supply is 240 volts. We need something approaching 100 volts to give us an electric shock, so that small batteries are completely safe. By the time we get to 240 volts, however, the shock that we get is so severe that it is quite likely to kill us by stopping the action of our heart muscles, as well as throwing us violently across the room as our arm and leg muscles contact. Surprisingly, when we slide our feet across a carpet and then touch a metal doorknob we may have actually charged ourselves up to more than 1000 volts, but the amount of electricity involved is so small that it is all discharged before any damage is done. In contrast, if we push a fork into an electric toaster and touch the heater element, all the generating capacity of the power station is available to pump electricity through our body. Household electricity must be treated with great respect.

While electric charge can pass through air as a spark, it needs a very high voltage to do so, and almost all the electric energy is dissipated in the spark. An extreme example is a lightning flash. The charge is separated in the violent air motions inside a storm cloud, generating a voltage of millions of volts. When the charge recombines, either from the top to the bottom of the cloud or from the bottom of the cloud to the ground, all the stored electrical energy is dissipated in the sound, light and heat of the lightning flash. This sort of gas discharge can be controlled and made useful in a neon sign or a fluorescent tube, in which the gas is at very low pressure, but air does not conduct electricity except under extreme circumstances.

It is practically useful to distinguish two sorts of materials on the basis of their electrical properties—conductors and insulators. All metals conduct electricity, though some such as copper, silver, gold and aluminium are particularly good conductors. Insulators are things such as glass, rubber and plastics. Some materials lie in between, such as salty water, or carbon, or the important materials called semiconductors from which transistors are made. Silicon is the most important semiconductor at present. It is made from sand or quartz, which consist of silicon dioxide or silica, and should not be confused with silicone, which is an oil or a rubbery plastic based on silicon and hydrogen.

Conductors of electricity contain free electrons—the "particles" of negative electric charge—rather in the same way as a water-pipe contains water, except that they cannot be emptied out. When we connect a wire to a battery or an electric generator and complete the circuit back to the other terminal, the electrons flow through the wire and back to the battery or generator. If we break the wire, then the flow stops immediately. In this way we can control the flow of electricity by using switches which simply break the circuit mechanically. This is really all there is to electric circuits—we can treat them rather like water mains except that none of the water must ever be allowed to escape. As a rough analogy, the electric current, which we measure in amperes (or just "amps") is like the total amount of water that flows past a given point in the pipe in one second, while the electric potential, measured in volts, is like the water pressure in the pipe. To do useful work we need a good flow of water at a good pressure,

Magnetism and Electricity

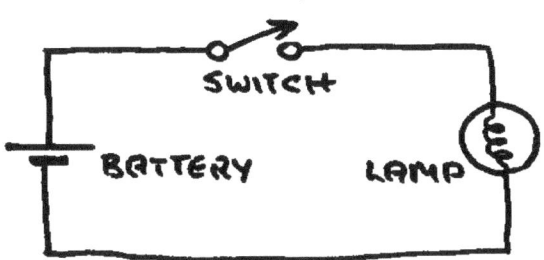

Electricity from a battery or from a generator must flow around a circuit of conductors and back to the battery or generator to do anything useful. We can therefore control electrical devices very easily by inserting a switch in the circuit which simply separates two pieces of metal and interrupts the circuit.

or in this case a good current of electricity at a good voltage. The power in watts is just the current in amps multiplied by the voltage.

The electrons do useful things simply by flowing round the circuit. They do not flow very fast—typically only millimetres per second—because they collide with the atoms of the metal and keep losing some energy to it, thus heating it up. If we make a small length of the wire very thin, then this must carry the same electric current as the thicker wire in the circuit, and consequently gets much hotter. This is almost all there is to an electric radiator, or indeed to a light bulb, in which the wire is very thin indeed. The only refinement we make is to use for the part of the wire that we want to heat up a rather different material that is not so good a conductor. This makes it get much hotter because it resists the flow of the electrons more. We also choose a material that will not melt too easily and, in the case of a light bulb, enclose it in a glass envelope filled with an inert gas (argon) so that it does not burn up, as it would in the air. In the meter boxes of our houses we always have a thin wire fuse (or some more sophisticated device) that will melt and cut off the electricity if the circuit is dangerously overloaded.

The other important interaction is that between electrons and magnetic fields. Michael Faraday (1791–1867) discovered that when a wire carrying an electric current is placed in a magnetic field then it experiences a sideways force. This phenomenon is the basis on which we can make an electric motor using a coil of wire in the field of a permanent magnet. Conversely, he also found that if we move a wire through a magnetic field then this generates an electric voltage between its ends—the basis of the electric generator, which is simply an electric motor in reverse. Without these two discoveries and the practical inventions that sprang from them, modern life would be very different indeed. An average home contains about 40 electric motors; a few of them are driven by batteries, but most are driven by electricity from a generator in a power station.

There is one important distinction between the electric current from a battery and that from a power station. A battery has no moving parts, of course, and the electric current simply emerges from one terminal of the battery, flows around the circuit, and returns to the other battery terminal. This is called a direct current (DC). In a power station, however, the electric current is produced by a coil of wire spinning in a magnetic field. Each time the coil spins once, electric current is forced a small distance in one direction and then sucked back again—we have and alternating current (AC). It might seem that an alternating current was of little use, because the electrons do not go anywhere, but remember that heating, for example, is caused by the electrons colliding with the atoms in the wire, and collisions can be made just as effectively by electrons moving backwards and forwards as by those moving in just one direction. Alternating current has advantages as far as transmission over long distances is concerned, and is now used almost universally. Australian power supplies are at "50 cycles" which means that the current goes through a complete cycle (two changes of direction) fifty times each second. We could also call this, more properly, "50 hertz".

When we look at electrical devices in the home we find that they usually have their rate of consumption of electricity marked on them in watts or kilowatts. The amount of electric power used is then determined by how long we have them running, and is measured in kilowatt-hours. A kilowatt is a good amount of power—it is more than the peak power that a strong human can produce for ten seconds or so before falling exhausted. It is more than enough to run a washing machine, and is just about the power of a single-bar radiator. A light bulb is generally about 60 watts, though of course there are more and less powerful globes available. Because a power station must supply tens of thousands of homes as well as hundreds of factories, it must have a very large power output. A typical modern coal-fired generating station in the Hunter Valley of New South Wales or in the LaTrobe Valley of Victoria has an electric power output of about 1000 megawatts, which is one million kilowatts.

So important is electric power that the prosperity of nations can be roughly measured by the amount of electrical energy used per inhabitant per year, or by the installed capacity of generating plant. In a country such as Australia, the electric power generation averages about one kilowatt per person for 24 hours per day, so that each of us has a powerful electric motor working for us somewhere in the country all the time. In many developing countries the available electric power is less than 100 watts per person, and in some countries not much more than 10 watts.

9

Atoms, Molecules and Chemistry

"Nothing exists except atoms and empty space; everything else is opinion." Democritus (460–370BC)

Suppose we have a rich Christmas cake and cut it into slices, then each slice is still recognisably Christmas cake. If we cut a slice into tiny pieces, however, then they are not all the same—some are pieces of nut, some pieces of date or other fruit, and a few are cake crumb with no fruit or nut at all—and none is recognisably Christmas cake. There is a scale of subdivision at which we have split the cake into its primary constituents.

Something similar happens with materials that are simpler and more uniform than Christmas cake, such as sugar or water for example. We can go on dividing these substances into smaller and smaller pieces until at the next stage we no longer have sugar or water but something different. This smallest characteristic constituent of a material is called a molecule. Molecules, we now know, are clusters of atoms rather tightly bound together, and atoms are, at least at the level of chemistry, the basic constituents of all matter. Indeed the word "a–tom" means "un–cuttable". These views are not particularly new—the notion of atoms goes back to the Greek philosopher Democritus (460–370BC)—but the chemical basis of modern atomic theory was laid by the Englishman John Dalton (1766–1844) only about two hundred years ago. In this chapter we shall summarise briefly what is known about the structure of matter at a molecular and atomic level.

Atoms and Elements

There are just 92 naturally occurring chemical atoms. This seems a strange number for the fundamental building blocks of chemical matter, but it comes about in a fairly straightforward way. After the "big bang" that started the universe on its expansion, most of the matter was in the form of hydrogen, the simplest and lightest atom. This hydrogen collected into galaxies and stars, as we have seen before, and in the interior of the stars the hydrogen was gradually converted to helium at immense temperatures. Along with this conversion, however, went many other nuclear reactions that built up heavier atoms still—oxygen and nitrogen and carbon and the rest, right out past uranium. As stars ran out of hydrogen, many of them exploded and scattered this wealth of chemical atoms throughout their galaxies where, as gas or dust, they later collected together to form planets orbiting round other stars. Many of the chemical elements were formed with more than one type of nuclear composition—they were isotopes of identical chemistry but with slightly differing masses. Some of these isotopes were radioactive and decayed away very rapidly so that they were soon gone; some were very nearly stable and decayed only over millions of years; some were actually completely stable. We are left now with only the stable atomic isotopes and with those few radioactive isotopes of extremely long lifetime. It just so happens that all isotopes of atoms with atomic number higher than 92 have short lifetimes, so that there are none of them left around. Uranium, which is element number 92 in order of mass, is simply the heaviest natural atom that is left. We can now make atoms up to number 100 or more in accelerators that mimic, in a way, the processes that occurred deep in the interior of stars, but they are radioactive and decay rather quickly to isotopes of smaller common atoms.

Materials made of just one kind of atom are called elements, and there are just 92 of them. Some of them are very important, particularly the metals, but their properties are generally quite simple and not very interesting. We can arrange all the atoms in a table—the Periodic Table—which groups them according to mass (or rather atomic number, which is not quite the same) and places them in rows and columns that demonstrate these chemical similarities. The members of some columns of the table, for instance, are all metals; those of another column are all inert gases; those of still another column corrosive gases; and so on. To simplify writing we use a shorthand symbol for each chemical element—H for hydrogen, O for oxygen, N for nitrogen, C for carbon, and so on. Of course there are only 26 letters in the alphabet, so we have to use two letters for many of the elements—Ca for calcium, Mg for magnesium, Cl for chlorine, and so on. Unfortunately, many of the elements were still known by their Latin names last century when the scheme was drawn up, so that we use the symbol Na (natrium) for sodium, K (kalium) for potassium, Ag (argentum) for silver, Au (aureum) for gold, and a few others likewise, but there are not many such exceptions.

Molecules

Much of elementary chemistry is concerned first with getting pure elements from naturally occurring minerals or other sources, and then finding out how they react with one another. Rather more than 100 years ago most of these simple reactions were worked out in detail. Atoms of some elements, such as silver and gold for example, simply mix together in any quantities without anything exciting happening, but when the two elements are very different, such as carbon and oxygen or hydrogen and oxygen, there is a reaction (burning in this case) which releases heat. It is found that there are simple rules for the amounts of the two materials combining in this way which can be interpreted to mean that these atoms form simple patterns when they combine—one atom of carbon always combines with two atoms of oxygen to form carbon dioxide, two atoms of hydrogen combine with one atom of oxygen to form hydrogen oxide or simple water, and so on. These new materials are called compounds, and the groupings of atoms that characterise the compound are called molecules. A molecule is the smallest part of a compound that still displays the properties of the compound.

Chemistry deals basically with molecules rather than with atoms, which are left to physics to worry about. Chemistry is about the reactions between substances, the building of complex molecules, and the properties of those molecules. Most of the simpler reactions have been known for a very long time, and have become a standard part of industrial chemistry—the making of acids, alkalis, fertilisers, dyes and so on. Modern chemistry deals in a much more subtle way with understanding biological molecules—which are often huge structures with hundreds of thousand of atoms—with designing new plastics and polymers, and with engineering antibiotics and other drugs at a molecular level.

There are so many different molecules in common use that we cannot hope to survey them or their properties, but it is interesting to have some pictorial idea of what they look like. We can get this from little schematic pictures of the arrangements of the atoms in the molecule as shown in the illustration. Water (H_2O) consists of two hydrogen atoms and one oxygen atom in a sort of boomerang shape, carbon dioxide (CO_2) has all three atoms in a straight

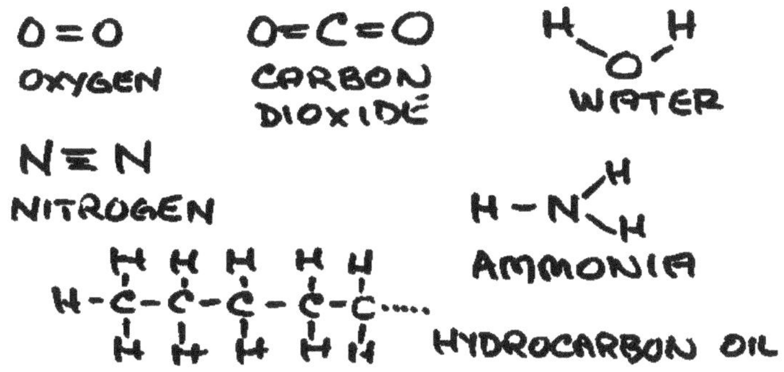

$O=O$ OXYGEN

$O=C=O$ CARBON DIOXIDE

H–O–H WATER

$N\equiv N$ NITROGEN

H–NH₂ AMMONIA

H-C-C-C-C-C.... HYDROCARBON OIL
(with H atoms above and below each C)

The shapes of molecules can be represented by simple pictures showing the arrangement of the atoms involved. There is, of course, some distortion because real molecules are not flat. The single and double lines connecting the atoms in a molecule have a precise meaning in terms of the way the atoms are bonded together, but this need not concern us here. In the molecules shown, H is hydrogen, O is oxygen, N is nitrogen and C is carbon.

line, methane (CH_4) has a tetrahedral arrangement which is hard to draw, so its is usually represented in a square. Biologically derived molecules such as oils consist mainly of carbon and hydrogen atoms, which is why they are called hydrocarbons, and tend to form long chains with carbon backbones. In more complex biological molecules, other groups of atoms hang off these backbones.

Crystals

Crystals are among the most fascinating non-living things found in nature. Most natural crystals are minerals that have formed in cavities in rocks—semi-precious stones such as the beautiful purple amethyst and more common once such as clear quartz—and are easily recognised by their shiny flat faces. Other natural crystals occur as inclusions in rocks—materials such as garnet, for example and shiny metallic minerals such as antimony or lead sulphide. Nearer to home, its is easy to see that both sugar and salt consist of tiny crystals. Careful study shows that the sizes and shapes of the crystals of one substance may vary, but what remain constant are the angles between the crystal faces. This is not

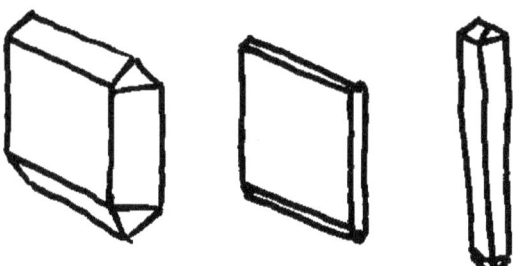

Crystals can have rather different shapes depending upon just how they are grown, but the angles between the faces of crystals of a given material are always exactly the same. This regularity is a reflection of the symmetry with which the atoms within the crystal are arranged.

very clear in the case of ordinary salt, which has often been ground up, but shows up very nicely in "coffee crystals" of sugar.

It is worth noting that precious gem stones such as diamond, sapphire and ruby are also crystals, though when found in nature they often have the appearance of rounded pebbles with dull surfaces because of having been abraded by sand and other gritty materials. It is only when cut and polished that they show their true beauty. Interestingly, the polished faces on such gem stones have no relation to the crystal faces that they would show in nature, and are purely decorative shapes arranged to reflect the light with proper sparkle.

Although it is the outer shape of crystals that brings them to our attention, the really important thing is that all the atoms that make them up are arranged in neat orderly rows, columns and sheets. The outer shape is simply evidence of this internal order, and nothing is changed if we polish the exterior to a spherical shape, for example. We can still tell that the interior is crystalline by careful experiments with x-rays.

It is a fascinating experiment to grow crystals of common materials that are readily soluble in water, such as salt or sugar. To do this we first have to make up a saturated solution by dissolving as much of the material as possible. Both salt and sugar dissolve more easily in hot than in cold water, so we usually start off with such a hot solution. As the solution cools down, and as the water evaporates with time, so the solution becomes more and more concentrated until the salt or sugar begins to precipitate out. We can either let nature take its course, in which case we will generally find a number of small crystals growing in clumps on the bottom of the container or, more interestingly, we can try hanging a small "seed" crystal of the material into the middle of the liquid on a fine thread. The salt or sugar then preferentially adds to the crystal already there, rather than starting again as small crystals, and we can grow quite a large crystal in a few days time.

A hand lens is all that is needed to examine crystal faces, and both salt and sugar are good starting points. We can easily see that the crystals of salt have faces that all meet at right angles, so that they are roughly the shape of bricks. Large crystals may appear to have more complicated shapes, but that is generally because several crystals have grown together. The shapes of sugar crystals are more complicated—some faces meet at right angles but some meet at other angles. This is evidence for the way in which the molecules are arranged inside the material.

While biological materials such as wood and bone clearly have an internal structure that is related to their origin and the way in which they grew, rather surprisingly most non-biological materials are actually crystalline. We can see this in rocks, particularly in the granite from which older bank buildings are often built, and crystals of several different minerals can usually be recognised. Ice is also a crystalline material, and we can see evidence for this in the six-sided shape of snowflakes. However, things such as the copper in water pipes and wires, the steel in gates and car engines, and the brass in door fittings are also crystalline. The reason that this is not apparent is that the metal crystals are very small—usually only a fraction of a millimetre across—and there are very many of them oriented in all sorts of directions. As we said before, it is the arrangement of the atoms that matters, not the external shape.

The one common non-biological solid material that is not crystalline is glass. Indeed the internal structure of glass is very much like the disordered structure of a liquid, except that all the molecules are fixed in place rather than being able to move. The difference between a glass and crystalline material shows up very clearly in the way they melt. Ice is an easy crystalline material to melt, and we see that it does so at a well defined temperature of 0°C and does not get soft or partly melt at lower temperatures. If we heat glass, however, it just gradually gets soft and turns into a treacle-like liquid over a large temperature range—there is no real melting point.

The study of the properties of crystals is an important part of physics, chemistry and geology. Part of this is because of the intrinsic interest of crystals themselves, and part because so many important things are made from crystalline materials. As we note above, all metals are crystalline, but the crystals are ordinarily very small. However when we want to make computer chips and similar devices we need very large perfect crystals of silicon, the base material from which they are made. Crystal growers can produce such crystals that are 15 cm across and nearly a metre long on a completely regular basis. Large crystals of sapphire are also regularly grown for small windows in high-temperature equipment, and even small diamonds can be grown in the laboratory, though these are not of gem quality.

Beyond Atoms

If molecules are composed of atoms, what are atoms composed of? A very reasonable questions, particularly remembering that 92 is not a very satisfactory number for the ultimate building blocks of the universe!

When Marie and Pierre Curie (1867–1934 and 1859–1906) began their study of radioactivity, they also began the era of investigation of the structure of atoms. Briefly, we now know that atoms are extremely small—about one tenthousand-millionth of a metre in diameter—and that they are nearly all empty space. At the centre of each atom is a tiny nucleus that accounts for most of its mass, and around this orbit electrons like a miniature solar system, the number of electrons, 1 to 92, corresponding to the atomic number of the atom concerned. The electrons are electrically negatively charged, while the nucleus has a balancing positive charge. Because of the large amount of empty space, atoms are a little bit "soft" in their interaction with each other, rather than being rigid balls, but the charges on the electrons keep them apart. The nucleus, for its part, consists of small heavy positively charged particles, protons, and small heavy neutral particles, neutrons. Because neutrons have no electric charge, nuclei can be induced to include one or two extra to form a different isotope, but this generally makes them unstable and radioactive. This model, with just three "elementary particles", provides a satisfactory sort of basis for the structure of matter.

Of course, the question then goes on—what are electrons and protons and neutrons made of? The answers are not yet completely clear, and they seem to complicate the picture rather than simplifying it. Briefly, electrons seem to be just themselves with no structure, but protons and neutrons appear to be made up of combinations of three other particles called quarks—a jocular name derived from James Joyce's obscure novel Finnegan's Wake. At the high energies

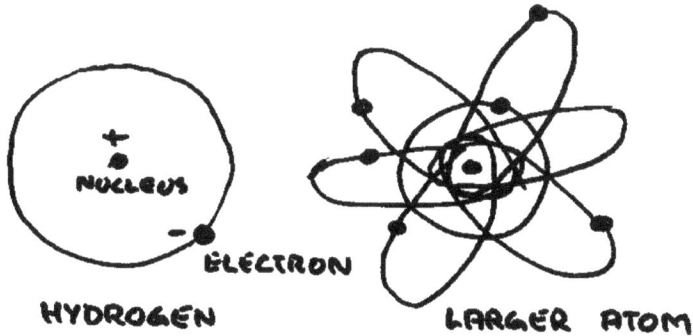

An atom consists of a tiny nucleus, containing nearly all the mass of the atom, around which orbit a sufficient number of electrons to just balance the positive electric charge on the nucleus. There are precise rules governing the orbits of the electrons, but they are to some extent "blurred out" in space.

at which the experiments are conducted there are also a host of other "particles" generated and it becomes unclear what one should regard as fundamental. For all practical purposes, a model with electrons, protons and neutrons is quite adequate.

Chemistry and Cooking

The world of chemistry is so rich and varied that it is difficult to survey it in any detail. What is probably better here is to look at a few common uses of chemicals and chemical knowledge and to show how the principles of chemistry can led us to an understanding of what is going on.

Food is required to provide nutrition for growth and energy for action, and that is the prime requirement in its choice. Fortunately humans are omnivorous, rather than being restricted to one type of food, so there is a good deal of choice in what we eat. Civilisation has also led us to expect more from our food than simple nutrition—it has an aesthetic as well as a practical dimension.

Much of our food is cooked, and one of the important things in cooking is to produce variety in texture and flavour. This applies particularly to food made from ground cereals and flour—a diet of porridge might be nutritious, but it would be very dull. Bread is an excellent example of variety in texture. Yeast cells are mixed with the bread dough and, as they reproduce using the sugar in the flour as food, they release carbon dioxide gas, which becomes trapped as little bubbles to make the bread rise. When cooked, this spongy texture is retained. The process of making bread is, however, a lengthy one, and a quicker method has been developed for sponge cake, damper, and similar things.

The essence of the method is to create numerous small bubbles throughout the initial liquid mixture, and there are two slightly different methods used. In the first the cook takes baking soda and mixes it into the other ingredients. Now baking soda is just sodium bicarbonate, a compound of sodium (common in salt) carbon and oxygen. When heated, this complex molecule decomposes, giving off carbon dioxide gas as desired and leaving behind sodium carbonate. This works well from a texture point of view, but the sodium carbonate ("washing soda") left behind has a slightly soapy taste which is noticeable if the other flavours are delicate.

The invention of "baking powder" solved this problem. This material is just a mixture the earlier sodium bicarbonate baking soda with citric acid, a weak acid derived from citrus fruit, and tartaric acid which is somewhat similar. Now when the cake mixture is heated the sodium bicarbonate decomposes as before, but the citric acid reacts with the sodium carbonate that remains, releases more carbon dioxide, and leaves a mixture of sodium citrate and sodium tartrate behind. These are materials with a very mild and quite pleasant taste.

It is not particularly easy to demonstrate this process directly, except by looking at the texture of the finished product. However the release of carbon dioxide bubbles can be shown by mixing the constituents in water. This works even better if one uses a rather stronger acid such as acetic acid in the form of vinegar and adds it to a solution of baking soda in water. The carbon dioxide that is evolved, incidentally, is just the same as that used to make carbonated drinks, but there the gas is dissolved directly in the sugary drink under pressure and is released as bubbles when the cap of the bottle is removed.

Incidentally, we mentioned the function of reproducing yeast cells in producing the carbon dioxide bubbles that make bread dough rise to a light and fluffy texture. Another major use of yeasts, of a rather different variety, is in the wine-making industry. Here the yeast cells feed on the sugars in grape juice, converting them to alcohol and releasing carbon dioxide. Something very similar happens in the brewing of beer.

Soaps, Detergents and Dirt

A problem that constantly faces us is to keep ourselves and our surroundings clean. Sometimes the problem is easy—the "dirt" from garden soil simply washes off in cold water, for example—and sometimes it is quite hard—a good smear of black oil takes hot water and soap to remove from our hands and even more prolonged treatment to remove from clothes. Some stains are simply permanent.

Modern industrial chemistry has provided us with a host of solvents, dyes and bleaches, some of which are important in household use. One of the best solvent of all is ordinary water, but of course most of the things that we find on the surface of the earth are those that are resistant to dissolution in water—the rest are already in the sea! One class of thing that does not dissolve in water is oils. The nearest thing we can get to dissolving an oil in water is to make an emulsion, which consists of tiny droplets of oil suspended in water. The most common emulsion in the house is milk. The largest fatty "cream" droplets tend to rise to the surface and form a layer, but the smaller oil droplets remain stably in suspension. To dissolve oils we generally need liquids that have carbon-based organic molecules like the oils themselves. Ordinary alcohol is a good example which works even when it is mixed with water, because its molecules also have a water-like end to them. We could use vodka, or some other colourless alcoholic drink, as a convenient solvent for oils or "clean" oily stains, but we usually have available methylated spirit, which is a mixture of ordinary (ethyl) alcohol with some methyl alcohol to make it unpalatable (and incidentally rather poisonous). Alcohol is also a good solvent for quite a number of dyes, and most food colourings are dissolved in an (ethyl) alcohol base.

When it comes to real dirt, however, the situation is different. Two major weapons in our fight against dirt are soaps and detergents. Somehow they act against certain sorts of greasy dirt—and that includes dirty skin—in a near-magical fashion. How do they do this? Actually soap is a very old friend, and is made by boiling animal fats with sodium or potassium hydroxide ("caustic soda" or "caustic potash"), while detergents are new synthetic products of the chemical industry, but they both act in the same way.

The molecules of a soap or a detergent are rather large and each has an oily end (from the fat in the case of a soap) and a polar or ionic end (from the caustic alkali). The oily ends of the molecules are happiest (that is, they have the lowest energy) when surrounded by other oily molecules, while the polar ends are happy when surrounded by water. This allows two things to happen. When we dissolve soap in water, we find that we can blow large bubbles which

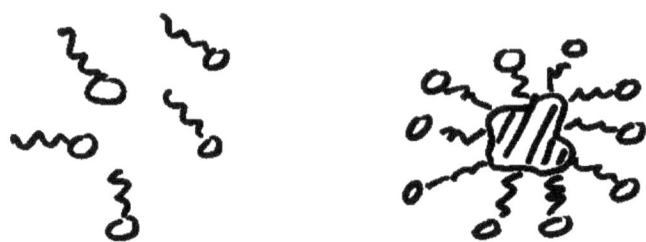

Molecules of soaps and detergents have polar ionic "heads" and long oily "tails". In contact with an oily dirt particle, these molecules orient themselves with their heads outwards towards the water, so that the dirt particle is easily dislodged.

persist for a long time. The reason is that the soap molecules can lower their energy by migrating to surfaces where they can have their polar heads buried in the water and their oily tails making a layer on the air-side of the surface. The thin water film of a bubble is thus protected by this soap-molecule film on both its surfaces. The second application, which is the one that is important to us here, is that, when soap molecules encounter a speck of oily dirt, they try to cluster around it with their oily tails touching it and their polar heads facing out into the surrounding water. A little bit of scrubbing lets the soap molecules cluster right around each speck of dislodged dirt and prevents it from sticking to the oily skin again, so that it floats off into the bulk of the water. Hot water helps this process because it melts the thicker oils and greases and makes it easier to dislodge the dirt particles, which are mostly sooty carbon when it comes to common oils.

Natural and Synthetic Chemicals

The chemical industry has been one of those most important to modern society for more than a hundred years. We can think of this industry under two headings—either it extracts complex chemicals as natural products from living things, or else it begins from simple raw materials and synthesises chemical products. It is important to realise that there is no difference between a "natural" product and the same chemical produced by synthesis, it is simply that in some cases one way of making the substance is easier than the other.

Simple bulk chemicals such as dyes, fertilisers, detergents and plastics are generally made by a synthetic route from simple chemicals recovered from minerals, coal or oil for the simple reason that this is the most economical way to do it. It does not make sense, however, to try to produce complex materials such as human foods by a synthetic route, because we are very critical about what we eat, and the direct natural route through growing plants and animals is quite efficient.

It is at the level of complex chemicals that are required in small quantities, such as drugs for medical use, that the most interesting things happen. As we see in a later chapter, many plants produce complex chemicals to attract insects to pollinate them and to repel other animals that might eat them. Not surprisingly, many of these chemicals have significant effects on humans if we eat them or if they enter our bloodstreams. In much the same way we can also extract active chemicals from certain parts of the bodies of animals that we kill to eat as food. Some of these plant and animal extracts are deadly poisons or have very undesirable effects, but others can be very useful in medicine. There is therefore a great deal of scientific interest in isolating such biologically active natural products and in testing how they affect humans, either to control body functions or to help cure disease.

Many traditional herbal remedies have been used in this way for centuries, and we now know just what is the active chemical involved. Aspirin, for example, the common headache remedy, was originally extracted from the bark of the willow tree, and quinine, a partial cure for malarial fever, from the bark of the cinchona tree. Many trees must be used, however, to produce not very much of the active ingredient, so that this natural route is an expensive one that makes it available to only a few people. More than this, the natural product can often come mixed with other plant extracts that have undesirable side effects. To make the benefits of these medicines available to as many people as possible, it has therefore been necessary to find ways of synthesising them from simple starting materials. Sometimes this is done by a direct chemical route, and sometimes by growing cultures of moulds or other life-forms in large reaction vessels.

The same route has been followed with many other natural pharmaceutical products, with the result that many drugs originally derived from plants can now be produced cheaply in pure form in the laboratory. Indeed, most of the expense of the pharmaceutical industry comes not from actually producing drugs but rather from the initial trial and testing phase to make sure that they are effective and that they have no undesirable side effects.

Today, organic chemistry has progressed to the stage that scientists are able to study viruses that cause disease and, after a great deal of effort, find the arrangement of molecules in the virus around the active site that allows it to multiply. It is then possible to design and synthesise a special drug molecule that will cling to the virus in such a way as to block its active site and make it harmless. In ways such as this, science is gradually building up an armory of drugs to combat all the more common viruses. Viruses, however, often mutate to slightly different forms, because they reproduce so rapidly, so that it is necessary to keep up a constant effort in the development of new pharmaceuticals to defeat them.

10

Useful Materials

"When the foeman bears his steel; Tarantara, tarantara; We uncomfortable feel; Tarantara." W.S. Gilbert "Pirates of Penzance"

Most things on the Earth can be used for something—some are good to eat, some are magnetic, some conduct electricity, and so on. In this chapter we concentrate on what are called structural materials—those that have properties such as high mechanical strength so that they are useful for making things out of. The number of materials we could think about is very large, so that it is helpful to group them into families with similar properties and then consider just a few representative examples of each.

Natural Materials

The earliest materials used by humans were those that they found ready to hand. Sticks and stones were first simply selected for size and shape, but later our ancestors learnt to chip stones and scrape wood to make implements. Some rocks, particularly the hard brittle material called flint, were particularly useful because they could be chipped carefully to produce a sharp edge useful for cutting. Other rocks came in convenient shapes and sizes for making the walls of simple houses, and were strong and heavy enough to provide protection from winds and wild animals.

Tree trunks and large branches provided structural materials that were long and thin, rather than being roughly round like a typical rock. They were also lighter than rock and so more easily handled. Such timber beams served to make the roofs of the stone-walled houses, and could be covered with dry grass or turf to make them fairly waterproof. Thinner straight tree branches were useful for hunting spears, and were often tipped with a sharp flint head to make them more effective. Australian aborigines even found a novel use for thin tree trunks hollowed out by ants, and made these into that unique musical instrument, the didjeridu (often spelt didgeridoo).

Finally, as a result of the action of fire, two other classes of material became available—the hard-baked earth, which later evolved into baked bricks for structures and into baked-clay ceramics for household utensils, and the simple metals copper, tin and lead that could be easily smelted from their ores by the heat of wood fires. The Roman armies came to Britain two thousand years ago, at least in part to take advantage of the tin ore that could be readily mined in Cornwall.

This was the beginning of materials science, and we still use these same structural materials for much of our building, although in carefully shaped forms. Master masons learnt how to cut and dress stone to carefully shaped blocks and used these to build the magnificent cathedrals of the middle ages—structures that would provide a challenge even to sophisticated modern building techniques. These were immensely high, with vaulted internal ceilings of interlocking stone, preserved from the weather by timber outer-roofs sheathed in lead. Timber was carefully selected for its end use, whether it was to make springy long-bows for the army, structural beams for castles, or decorative furniture. Readily extracted metals such as copper and tin were combined to make strong bronze for weapons and for household utensils, while lead was rolled up to make pipes for early water supplies.

Ceramics

Ceramics made by softening clay with water, shaping it, and firing in a wood fire have been made for something like 10,000 years, and the fact that they can be recovered almost undamaged by time from the ruins of ancient civilisations shows how well the potters knew their craft. Baked clay bricks have lasted similarly in buildings and, even more importantly, the clay tablets on which scribes wrote important inscriptions and then baked them to ensure permanence.

While the process is simple, the details of what happens are less so. The water-softened clay is somewhere between a very stiff liquid and an easily-deformed solid in its properties, which makes it ideal for molding. The heat of the kiln in which it is baked drives off the water and then causes the particles of clay or other material to stick rigidly together in a process called sintering at a temperature that is well below their melting point. The glaze that is often applied to the surface, in contrast, consists of ground glass and colouring minerals, and melts completely to form a smooth coating.

Ceramic articles for domestic use are still made in almost the same way today as they were thousands of years ago. All that has changed is the use of finer clay and the application of glass-based glazes, themselves of ancient origin, to protect and beautify the surface. Automated processes have replaced the potter's wheel for producing cheap mass-produced ceramic teacups and the like, but craft potters maintain the ancient traditions.

Ceramics also have an important place in more advanced applications. Electric insulators are made from baked glazed ceramics, the magnets of loudspeakers and those used to hold notes onto fridge doors are generally iron oxide ceramics with magnetic properties, and new sophisticated ceramics such as zirconia, made from special beach sands, are used in wire-drawing dies and in parts of diesel engines.

Metals

The most important development of ancient materials technology is, however, the exploitation of metals for structural uses. The early metals recovered using wood fires were copper and tin, but both of these were too soft to be much use—feel how easy it is to bend a copper wire, and pure tin is rather similar. Adding about 10 percent of tin to copper, however, produced the alloy bronze that is very much harder than either copper or tin. So important did bronze become for weapons and utensils that the age of its discovery has become known as the Bronze Age.

The smelting of iron ores to produce metallic iron required higher temperatures and became easy only with the development of the blast furnace, in which a blast of air was blown through the burning coke heating the furnace. The material produced by the furnace is cast iron or pig iron, which is strong but brittle because of the high carbon content from the coke. Steel, which has had the carbon content reduced to less than 1 percent, is much tougher, and can have its hardness controlled by heat treatment. If still more carbon is removed we get mild steel, which is very generally useful but much easier to bend, while if all the carbon is removed then the pure iron is too soft for most uses.

Steel is stronger, more plentiful and cheaper than copper, tin or bronze and has become the basic structural metal of the twentieth century—iron or steel bridges, building girders, roofing sheets, railroads, vehicles, tools, engines ... The list is nearly endless. Its one disadvantage is that it is quite heavy—about eight times the weight of an equal volume of water—so that it is not a good material for use in aircraft or other places where weight must be kept as low as possible.

The twentieth century has seen the development of light alloys based on aluminium for use in such applications. Aluminium can be recovered from ores such as bauxite, but its production uses a lot of electrical energy so that it is much more expensive than steel. Its density (mass per unit volume) is less than three times that of water and, though it is not as strong as steel, it can be given extra strength and hardness by adding about ten percent of copper and small amounts of other materials. Even lighter alloys based on magnesium, which is less than twice the density of water, have also been developed and used in aircraft.

When choosing a metal for some structural use we must consider not only its strength, weight and appearance, but also its cost and how long it will last. Materials engineering, like all practical applications of knowledge, is a matter of balancing advantages against disadvantages to achieve the best practical solution.

Plastics

Plastics do not occur in nature but are manufactured from simple ingredients by using sophisticated chemistry. They bear some resemblance to biological materials such as skin, hair and fingernails and, like them, are made essentially from atoms of carbon, oxygen and hydrogen linked together to form molecules in the shape of long chains. The essential thing about a plastic is that a further step then takes place in which the molecules themselves are cross-linked together to form a coherent mass called a polymer (*poly* meaning many and *meros* meaning share). Sometimes this cross-linking is accomplished simply by adding another chemical and stirring, sometimes it is done by heating, sometimes it is an integral part of the manufacture.

Because plastics are made rather than found, they can be engineered to have many different properties. Some are hard and brittle, some are soft and flexible; some are transparent, some are opaque, many are coloured. Because they are made from small atoms, they are all relatively light. They can be manufactured as rods or sheets or filaments of any length and thickness, or they can be molded directly into the shape of the finished article.

Some of the most important and best known plastic materials are nylon (fine threads or small molded articles), polyethylene and polystyrene (usually molded articles), though there are thousands more. Most polymer plastics can be recycled into other less-refined plastics and used for things such as garbage bins, though it is necessary to go by different paths depending on the starting materials. This is why recyclable plastics are identified by numbers so that they can be placed in the correct bins. It is important in most applications, such as motor vehicle body parts and plastic containers, that the material last well, and this later causes a problem if these objects are simply thrown away rather than being recycled.

Foams and Fabrics

In many applications, but particularly in clothing and upholstery, we need materials that are soft and durable and that contain a large amount of air. The reason for this is that air, if it is trapped in bubbles or other small spaces, is a very poor conductor of heat. Thick fluffy fabrics and plastic foams are thus excellent thermal insulators and we can use them to keep hot things warm and cold things cool. Plastic foam picnic boxes are a good example. In the case of clothing, the aim is to help control the loss of heat from the human body, where it is generated by digesting food, to the air, which is generally at a temperature lower than that of the blood ($36.9°C$).

The effectiveness with which this can be done depends upon the amount of trapped air and the size of the pores. In the case of fabrics, good insulation properties suggest that we should use very fine fibres and not weave them too tightly, giving a fine soft fabric. At the other end of the scale, if we use coarse nylon filaments woven into a tight mesh, then the resulting cloth may be unsatisfactory as a clothing fabric, but may be excellent for making soft travel goods, fabric containers, or protective clothing.

Materials and Design

Materials and design interact very strongly. The mechanical properties of available materials determine the necessary size and thickness of elements of the structure, and this flows on to the whole design. A bridge built of stone will necessarily look very different in design from one built of cast iron, such as many of the early railway bridges in Britain, and this will in turn look very different from one built of high-tensile steel or of steel-reinforced concrete. A child's toy made of wood will look very different from the same sort of toy built from metal sheet or from moulded plastic.

The essence of good design is to choose the most appropriate material for the purpose on hand, and then to arrange the design so as not only to exploit the strengths and minimise the liabilities of the material, but also to exhibit something important about the relation between material and design. Metal has strength and surface sheen and can be made in very long uniform lengths, stone has surface texture and solidity and weight, silk has sheen and drape, wool has softness, timber has depth of colour and individuality. There is, or should be, a very close aesthetic relationship between objects or structures and the materials from which they are made.

11

Life and Living

"We are such stuff as dreams are made on, and our little life is rounded with a sleep." Shakespeare "The Tempest"

From our point of view, life is the most important thing in the universe. Certainly it is the most unexpected and complex phenomenon of which we are aware, and we continue to find out more about it. It poses fundamental questions: How did life and living things come about? Is life on earth unique, or is there life somewhere else in the universe? What is the place of human beings in the realm of life? What is consciousness? and so on. We cannot hope to answer any of these questions fully here, and most of them we will have to simply defer to another day, but it is important to realise the complexity of the whole life phenomenon.

First then, we should really ask "What is life?". This is a very serious question, but one to which we cannot hope to have a proper scientific answer—it is not really a question of science, but a question of the meaning of words. There is no agreed definition of what the word "life" means, although of course we can list many of the attributes of life, so that we cannot really hope to answer the question. Quite apart from this, it actually seems that there is no sensible point at which one can draw a dividing line between living and nonliving or inanimate. A bacterium is certainly "living" on the basis of any sensible definition of the word, while a sugar crystal is nonliving, even though it is biologically generated and able to grow from solution. But how about a virus, which has some of the attributes of a bacterium and some of those of a crystal? Perhaps we can come to a decision on the basis of whether or not it contains the genetic material DNA, but that seems too chemical a criterion, because we then have to ask whether a DNA molecule is living. The answer is that it doesn't really matter. If we decide one way or the other for the virus—and "living" seems to be a reasonable answer—then we will likely find another point of controversy somewhere between the virus and the sugar crystal.

Quite apart from these difficulties about the meaning of language, biology encounters another problem which looms large in many people's minds. Biology, as a science, functions at two levels. The first examines the operation of processes in living organisms; it builds theories and models of the way these systems operate, and is easily tested by repeated experiments as in the physical sciences. The second part of biology tries to provide a coherent explanation of the way things came to be like this—the origins of life and of its huge variety. Here experiments are mostly out of the question, and one can only test the models for logical consistency and for conformity with the facts and general principles established in both biology and other branches of science such as physics and geology.

The Origin of Life

As we discussed in Chapters 5 and 6, the universe has been around for a very long time—about 15,000 million years—and even the Earth has been solid for some 4500 million years. That gives a hugely long time for things to happen!

As the Earth cooled from its molten state, the immense amount of water vapour condensed to form oceans and shallow seas. As well as water, there was a large supply of other simple molecules such as nitrogen, carbon dioxide and methane, and the high temperatures meant high evaporation and high rainfall. Violent thunderstorms were probably very common, volcanic eruptions sent new materials into the oceans, and the whole surface of the Earth was in a state of flux. What is more, this state of things went on for one or two thousand million years. In that time, anything that could reasonably happen from a physical point of view was quite likely to occur. In particular, although chemical forces tended to break things down, they also allowed the building up of complex molecules of all kinds. There is direct evidence that this does occur. In laboratory experiments, flasks containing just simple molecules such as methane, carbon dioxide, ammonia and water, known to have been present on the Earth's surface, have been warmed and shaken for periods of months, with occasional electric discharges passed through them to simulate the effects of lightning. When they were finally opened and analysed, traces of complex molecules, including simple amino acids which are the building blocks of living organisms, were found. And that was after only months, not after a thousand

$$O=O \quad \text{OXYGEN}$$
$$O=C=O \quad \text{CARBON DIOXIDE}$$
$$H\text{-}O\text{-}H \quad \text{WATER}$$
$$N\equiv N \quad \text{NITROGEN}$$
$$H\text{-}NH_2 \quad \text{AMMONIA}$$
$$H\text{-}C(H)(H)\text{-}C(H)(H)\text{-}C(H)(H)\text{-}C(H)(H)\text{-}C(H)(H)\ldots \quad \text{HYDROCARBON OIL}$$

Complex organic molecules such as amino acids can be produced simply by random chemical actions in a mixture of common molecules ordinarily present in the atmosphere, provided a sufficiently active energy source is available.

million years!

Of course, simply to make molecules that are the constituents of living things is by no means enough—some agency has to be found to assemble them into much more complicated structures. We don't yet know how this happened, but some interesting clues are beginning to emerge. Indeed it seems to be a feature of the way in which physical principles operate that order tends to emerge out of disorder. The random swirling of winds in the atmosphere organises itself into giant pressure systems that sweep across the surface of the Earth; solutions precipitate to produce exquisitely ordered arrangements of atoms in crystals; and even the apparently random motion of a pendulum above a hidden magnet shows a deep and subtle order.

All this is a long way from the origin of life, but we can see similar things at the molecular level. Many simple molecules containing just carbon, oxygen and hydrogen have a polar head and a long oily tail. Here "polar" just means water-like, and refers to the part of the molecule containing an oxygen atom bonded to a hydrogen atom, while "oily" refers to the part of the molecule that is just carbons and hydrogens. These molecules can undergo a spontaneous self-organisation that is very relevant to life processes. The polar heads of the molecules are happiest (which means that their energy is lowest) if they are surrounded by water molecules, while the oily tails are content if they keep as

 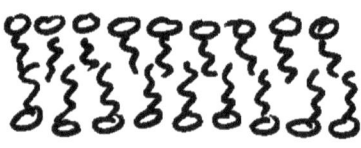

Molecules with a polar head and an oily tail can self-assemble into simple globular structures called vesicles, or into double-layer membrane structures.

far away from water as possible. There are two ways in which they can achieve this, simply by random trial-and-error processes. In the first structure, called a vesicle, the molecules clump together into tiny spheres with their oily tails coiled up inside and the polar heads forming the surface. The second arrangement is a little more complicated, and involves the molecules arranging themselves into two sheets, in each of which all the tails point in one direction. If these two sheets then come together with their oily sides in contact, they form a membrane with only polar heads exposed to the surrounding water.

Although vesicles are not particularly important to our story, membranes are, because a membrane can close up to enclose a small volume of watery liquid and protect it from outside influences. This is just what happens in a living cell, and the membranes that protect it are of just the kind we have described. More than this, if there are other more complex molecules available with similar oily and polar parts, they may be incorporated in the membrane to give it special properties, such as letting through certain small molecules and excluding others. In this way the interior of a membrane cell can develop by purely chemical means into a rather special environment.

Again, we are a long way from life itself, but some of the structural elements are there and can be seen to have arisen quite naturally. The next step is more complicated and we have as yet no idea how it occurred. But the simplest living things consists just of a single cell protected by a membrane and containing particular large complex molecules called nucleic acids. In some types of cells these nucleic acid molecules simply float round inside the cell, but in other types of cell they collect together into a tiny nucleus that is itself surrounded by a membrane. These molecules of genetic material are the really "living" part of the cell. The most important nucleic acid is the huge molecule DNA (deoxyribonucleic acid) which contains all the genetic information about the structure of the cell. A vitally important clue to the unity of all life on Earth is the fact that the DNA molecule is common to all forms of life—plants, insects and human beings—though details of the structure of the molecule vary from one species to another. We suspect that DNA may have evolved by some sort of self-assembly mechanism like the cell wall, but what we do know is that is can make copies of itself by a very neat molecular mechanism.

The DNA molecule has a double-helix structure, the determination of which won the Nobel Prize for Francis Crick, James Watson and Maurice Wilkins in 1962. Each strand of the helix is made from amino acids, of the sort that can be made in the elementary ocean soup with the help of electric discharges, and the links of the two strands are exactly paired. The molecule can replicate itself in an environment rich in the appropriate amino acids by separating into two single helices, each of which then builds up into a double helix again using amino acid building blocks from the surrounding liquid.

In this way the genetic material of a cell can duplicate itself exactly. The two sets of genetic material, whether organised into nuclei or not, then move to opposite ends of the cell and the membrane develops a waist between them and then separates into two separate cells. The whole process is called mitosis, and is the way in which living things grow. The very simplest living things consist of only a single cell, and after the duplication process the two cells separate and drift apart. More complex animals, such as human beings, consist of huge numbers of different cells which continually split and multiply in this way. Some parts of the body, such as hair and fingernails, simply grow longer until they are worn off or cut off, but in most parts of the adult animal the new cells replace other older cells that have died.

The DNA genetic material in the nucleus of a cell determines its properties and its reproductive behaviour. So long as the DNA copying process proceeds without errors, the two daughter cells are exact replicas of their common parent, which has now disappeared into them. But of course errors can happen in the process—the DNA strand to be copied may involve assembly of hundreds of thousands of amino acid building blocks. There are several possibilities if a copying error, or mutation, occurs. Either it will be serious, and the daughter cell with the error will fail to function properly and die, or it will be less serious

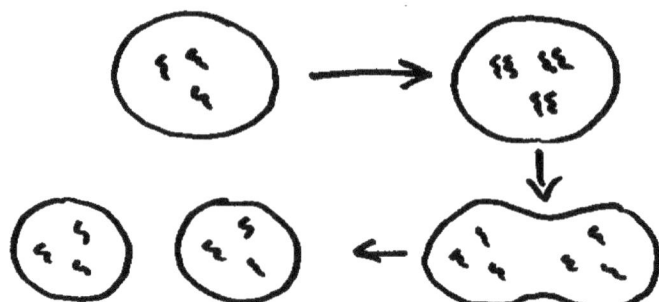

In multiplication of a simple cell by mitosis to form two identical daughter cells, the genetic material first replicates itself, the replicas separate, and the cell divides.

and the cell will live and divide again in turn.

If the altered daughter cell does survive, then again there are two possibilities. It may be nearly identical to its parent—perhaps a little larger, or a little more vigorous, or of a slightly different shape—and then no harm is done. Sometimes, however, the variation is more extreme and the new daughter cell may be radically different in its properties. If it is part of a multi-celled organism such as an animal, then this radically different cell may present grave problems, particularly if it reproduces vigorously. The new cells its produces will be like itself and will be unable to fit in with the part of the animal to which they properly belong. They will form a tumor or even a cancer. We do not know a great deal yet about what influences the formation of such tumor cells. Certainly part of the problem may lie in the original genetic material, which may encourage cells with errors to grow rather than to die, but also certainly there are chemical substances—carcinogens such as the tar in tobacco—which interfere with DNA copying and promote the formation of defective cells.

The Variety of Life

Populations of cells that reproduce by mitosis as described above can change slowly by random non-lethal copying errors, but otherwise remain fairly stable over time. Another form of cell reproduction has evolved, however, which allows

> In sexual reproduction, called meiosis, two germ cells each with only half
> the normal amount of genetic material combine to form a normal cell. This
> cell can then reproduce by duplicating its genetic material and splitting
> by mitosis.

mixing of genetic materials from cell populations that are very slightly different. This is called meiosis, and is the basis of sexual reproduction. Cells that reproduce in this way divide so that, instead of producing daughter cells with a nucleus that is an exact copy of the parent nucleus, the daughter cell contains only half of the genetic material. Daughter cells from a single individual do not generally recombine to form whole cells, but daughter cells from different individuals can do so, effectively mixing the genetic material of their two parents. As we see later, this has some particular advantages.

In most multi-cell organisms such as plants and animals, only cells in one particular part split in this way, while all other cells reproduce by mitosis. The special organs that produce cells with only half the genetic material are called reproductive organs, and we might call the cells they produce germ cells. Some plants have two sorts of reproductive organs, one of which releases germ cells to fertilise other individuals, while the other type is to catch germ cells and combine them with its own germ cells. In plants, the first type of cell is represented by the pollen produced in flowers and the second by the base of the flower which develops into fruit. Most animals, however, and some plants, divide into two sexes with one type of reproductive organ each. Male reproductive organs produces sperm cell which are available for fertilisation, while female organs produce eggs which are the receptors.

The particular feature of sexual reproduction is that it allows considerable variation between individuals, while maintaining the integrity of the species as a whole. The genetic material contributed by the male and the female is simply half of the available stock in a cell, and it can be divided up in many ways. Rules are known for this, but we won't go into details here.

Now consider what happens in a population of individuals over time. There will be a random variation, within limits, because of the random combination of genetic material in reproduction. Lethal variations will be quickly eliminated, but others will persist. If a variation confers no particular advantage, then its persistence will be random. Some variations, however, may confer an appreciable advantage, such as producing a few more offspring or needing a little less water, or having sharper teeth. Over time, individuals with these characteristics will prosper a little more than those without them, and the genetic makeup of the population will drift. The time required will depend upon the variation involved, but one might estimate perhaps a hundred generations as being reasonable. This amounts to around a thousand years for a large animal, but can be only a matter of months for a microbe.

Of course, with the aid of purposeful intelligence, the random element can be eliminated and genetic shifts can be made much more quickly. Plant breeders can produce new varieties in times of around ten years, while breeders of domestic animals such as dogs take rather longer. In these two cases, however, the new variety would rarely be successful in nature. Seedless fruits, for example, are highly prized for eating, but depend upon the presence of gardeners to propagate them as cuttings if they are to survive.

The general premise of the theory of evolution of species, as put forward by Charles Darwin (1809–1882) is just this—random variations in the genetics of populations that confer an advantage on their possessors tend to dominate, while those that are unfavourable are suppressed. One could therefore suppose that, over the thousands of millions of years that there has been life on Earth, there has been time for isolated populations of plants and animals to diverge quite markedly from their common ancestors, while retaining useful family features. We look at some of those common features and some interesting variations in the next two chapters.

The general course of evolution is influenced, naturally, by changes in the environment. If lakes dry up over ten million years, then there is a chance for aquatic creatures to adapt their gills to extract oxygen directly from air, and those populations that can do so survive. Even a small adaptation in this direction confers an advantage. Scales can similarly turn gradually to lighter feathers, and feather colours can change to blend with the environment. Even a

Life and Living   101

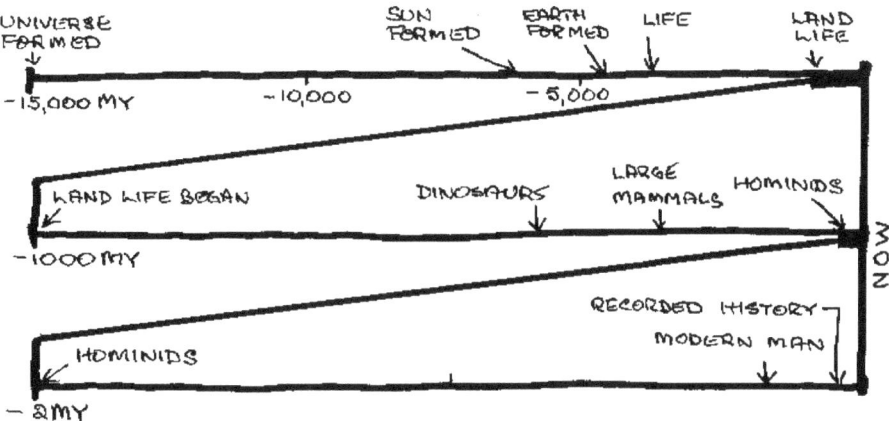

The timescale over which life has developed on the Earth is immensely
long. This progressively-expanded chart gives some idea of scale.

complex organ such as the human eye can be seen to be the result of progressive
evolutionary adaptations. A generalised sensitivity to heat and light on one
patch of skin provides an advantage to its possessor; if the sensitive skin is in a
fold, then its sensitivity can increase; if the fold is covered over by thin transparent skin and filled with liquid, then there is another increase in sensitivity,
and so on, all these adaptations being of competitive advantage to the group
possessing them. At this stage, some eyes went one way and some another,
giving the different eyes of insects and of vertebrates.

Trying to trace the complex history of the evolution of life on the Earth is a
very difficult matter, and the traditional methods of science are of only limited
help because much of the evidence is indirect. Various evidence suggests that
life began on Earth about 3500 million years ago—not much more than 1000
million years after the Earth itself became solid. The primitive plant life forms
that existed at that time were probably responsible for generating all the oxygen
in the atmosphere on which more modern life depends.

The occurrence of fossils in rock beds gives us much more detailed information about more recent, but still very ancient, forms of life. Fortunately fossils
are widespread and cover a very long period of geological time. The earliest
fossils, mostly simple shells and parts of plants, date back to about 590 million

years ago. This is a very long time ago, but already something like 3000 million years after the beginnings of the primitive life responsible for the atmosphere. Some species, such as simple shellfish, have remained almost unaltered up to the present day, but mostly the history revealed by the fossil record is one of increasing diversification, with different species being dominant at different times. Some whole families of plants or animals fail to survive a major environmental change. The dinosaurs, for example, dominated the Earth from 200 million years ago to 50 million years ago, but became extinct quite suddenly (if you say that extinction over a few million years is sudden!) probably as a result of environmental change, with that environmental change possibly due to a large meteor impact. Their disappearance left room for the development of larger mammals and ultimately of primates and human beings.

Animals that are recognisably human and different from primates such as monkeys have lived on Earth for about 2 million years, judging from the fossil record. Modern man, *homo sapiens*, is recognisable over only the past two hundred thousand years or so, though there is a continuity of development from earlier *homo* species. Some people are distressed at the thought that humans are animals, very like other animals, and have evolved to their present state by an extremely long process. To most of us, however, this realisation is a comfort. We are one with our animal cousins but, having evolved much more capable brains, we can appreciate how we came to be what we are. We can recognise our powers over the environment, and we must also recognise our responsibilities.

These huge time scales are a little difficult to comprehend, and it may be helpful to consider the whole life of the universe up to present time as compressed into one school day, from 9AM to 3.30PM. On this scale, the "big bang" that is thought to have signalled the birth of the universe occurs as the 9AM bell rings. By morning recess most of the galaxies and stars have formed, and planets are beginning to condense from the clouds of dust and gas that surround them. Soon after lunch the Earth is in a stable orbit and its surface has cooled sufficiently to solidify—this is the beginning of geological time. Life of a primitive sort begins to populate the oceans soon after—by about 1.30PM—but it is not until after 3PM that animals begin to emerge onto the land. By about 3.20 the Earth is covered by dense forests that will ultimately form coal beds and oil reservoirs. Dinosaurs are the dominant large lifeforms, but they rule the earth for not much longer than five minutes before their mass extinction at about 3.28PM. Now is the time for the mammals, but primitive hominids do not emerge until it is only about five seconds before 3.30. Modern humans, *homo sapiens*, enjoy the final eye blink of less than half a second before the bell signals that we have arrived at the present time.

Life and Living

It is only with the realisation of this immense sweep of time that physical and biological evolution become reasonable propositions and humanity falls into its proper place at the end of an almost inconceivably long chain of events. It is fascinating to speculate on the possibility—even the probability—that life in some form or other has developed on many planets circling many stars in the universe. The physical conditions for this to occur certainly exist on millions of planets, and our understanding of life on Earth suggests that self-replicating molecular systems tend to develop naturally, given even remotely suitable conditions of temperature and chemical environment.

If some form of life exists on other planets somewhere in the universe, then it is just as likely to be more advanced than our own human life—corresponding to perhaps 3.31 on the school clock and thus millions of years on in our own evolution— as it is to be more primitive. But the immense distances of space make it unlikely that we shall ever meet. Even to exchange signals by light or radio takes eight years to the nearest star, and perhaps hundreds or even thousand of years to the nearest populated planet. Our present understanding suggests that this is all we can do, but it is part of the fascination of science that new possibilities continually arise. Who knows what may ultimately be possible?

12

The World of Plants

"I think that I shall never see A poem lovely as a tree." Joyce Kilmer

There is a fairly clear distinction in most people's minds between plants and animals—plants are rooted in place, green and harmless, while animals run around, eat things, and are sometimes dangerous. Television nature programs have shown some interesting intermediates, such as the Venus Flytrap plant, which traps insects and dissolves them, but the distinction seems a pretty good one. This division is a useful one from a scientific viewpoint too, but requires some refinement, and even then there are organisms that do not fall easily into one class or the other. This is of no particular concern to scientists, because there is no reason why nature should conform to the rather rough and arbitrary conventions of human language, and classification is simply a convenient means of organising knowledge rather than anything fundamental to science. The classification of species of living things is an important branch of biological science called taxonomy.

Broadly speaking, plants are living organisms that grow by taking in simple chemical substances and building them into complex molecules with the help of energy from the sun, while animals are living organisms that grow by taking in food consisting of complex molecules and breaking them down without the need for sunlight. Plants mostly do not move around, but are tethered to one place by some sort of attachment, while animals can generally move from one place to another. At the level of large plants and animals there is usually no difficulty in deciding whether a specimen is an animal or a plant, but for very small and often single-celled organisms things are less simple. Indeed it is a useful way around the classification problem to introduce another category called "microorganisms", the members of which are often not easily classified

The World of Plants   105

as either plants or animals. We have written as much as planned about single-celled organisms in Chapter 11; here we concentrate on larger and more complex living things.

Photosynthesis

The structure of plant cells, and the ways in which they reproduce, are very similar to those described for all cells in Chapter 11. Apart from peculiarities of shape, the main feature of plant cells is that the cell wall is made up from the complex molecule cellulose. The real distinction, however, is in the way the cells get food and energy for their growth and reproduction. The process involved is called photosynthesis.

The materials needed for cell growth are carbon, oxygen, nitrogen and hydrogen atoms, with just traces of other materials, and these are all ready to hand. The air consists of about 78 percent of nitrogen, 21 percent of oxygen and about 1 percent of the inert gas argon. It also contains a small and variable amount of water (hydrogen dioxide) and a tiny amount (about 300 parts per million or 0.03%) of carbon dioxide. The green substance in plants is the complex molecule chlorophyll, which absorbs the red and the blue parts of the sun's energy and converts the carbon dioxide and water available from the air, or from the soil through its roots, into sugar. This sugar is then the food from which the plant gains all its needs.

The process of photosynthesis has very important consequences for life on Earth, because when sugar is synthesised from carbon dioxide and water by the chlorophyll, there is oxygen left over and this is released into the atmosphere. It was the action of this process in primitive plants over immense periods of time that provided all the oxygen that is now in the atmosphere and upon which animals depend for survival.

Sugar is a rather simple molecule, and plants need to modify it chemically to build up the cellulose, amino acids and other materials that they need for cell growth. This modification of the sugar, to provide energy and large molecules for the plant growth process, uses up oxygen, and releases carbon dioxide—just the opposite of the photosynthesis process. This reverse process goes on all the time, but in the light of day the photosynthesis process produces a great deal more oxygen than the plant uses. In the night, however, there is no photosynthesis, and the plant absorbs oxygen and releases carbon dioxide just like an animal.

We mentioned nitrogen and various trace elements as also being essential needs for plants—nitrogen, for example, is required to make the chlorophyll

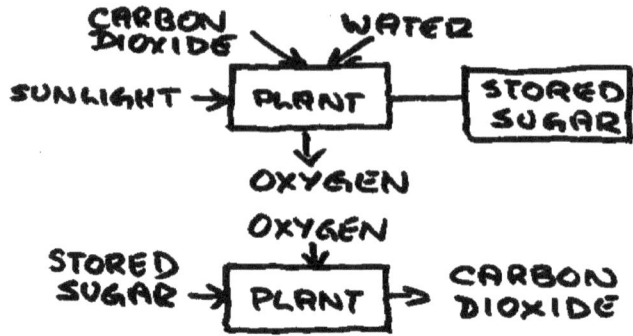

In sunlight, plants combine carbon dioxide and water to give sugar and release oxygen by the process of photosynthesis. At the same time there is a constant slow process of sugar use which consumes oxygen and releases carbon dioxide. In the dark this second process is all that happens.

molecule—and these are mostly provided from the soil through the roots. Plants are not able to use the nitrogen in the air directly, but rely upon bacteria living on their roots or in the soil to produce soluble nitrogen compounds. The fertilisers that we apply to plants are generally rich in nitrogen compounds, as well as containing phosphorus, iron, and other essential minor components for making plant cells.

Water is in a rather different category from the other essential materials for plants, because it is needed not just to supply chemicals for the photosynthesis process, but also as a filling for all the plant cells and as a transport medium within the whole plant. There is not usually enough moisture in the air to supply this water demand, and a major task of the roots is to extract water from the soil.

How Plants Live

Because plants are living things, they are engaged in a perpetual struggle to grow, to discourage predators and to reproduce, a struggle that is no less real than those of the animal world. The structure of a typical plant divides up these functions.

The leaves have large surface area to catch the sunlight, and contain a high concentration of green chlorophyll for the photosynthesis process. They have tiny pores (stomata) to allow in the essential carbon dioxide and to allow out the oxygen—or movement in the opposite direction at night—and a system of water-filled veins to allow transport of the manufactured sugars to other parts of the plant. These leaf veins have to be carefully arranged to reach all parts of the leaf, but must not allow the escape of too much precious water. Because the leaves are soft and nourishing to many animals, plants often manufacture poisons to make their leaves taste unpleasant and to discourage, if not actually killing, these predators. Humans have learnt to circumvent this mechanism by selectively breeding plants without unpleasant poisons, and repaying them by protecting them with insecticides, caring for their seeds, and generally fostering their growth and reproduction. Some ants and other insects have developed a similar symbiotic relationship with particular plant species over the ages.

The trunk and branches support the leaves and may contain the main bulk of the plant, although this varies greatly from one species to another. They also provide the connecting pathways for the transfer of nutrients. The trunk or stalk, in particular, contains vertical channels through which the watery sap containing the nutrients flows, surrounded by a mechanically tough outer skin or bark. Because the trunks of large trees have to be strong to support their branches, they have developed cell structures appropriate to this purpose by the processes of natural selection. At the same time they have provided a source of strong construction materials for the use of predators such as human beings.

The roots, too, have a dual role. On the one hand they are the organs by which the plant collects water, soluble nitrogen and trace elements from the soil; on the other, the roots provide the mechanical anchorage for the plant against the stresses of wind and water and the shear weight of the leaf canopy. The roots also provide an environment in which microorganisms such as fungi can flourish, often providing nutrients in useful form for the plant.

Reproduction

The flowers and fruit of plants are so familiar that they require little discussion. The pollen cells carry half of the genetic material of a plant and the seed cells the other half, fertilisation taking place either by wind-borne pollen dispersal or with the help of insects attracted to feed on the nectar supplied by the flowers. Seeds develop in the fruit, and again there are many methods of seed dispersal. Some seeds are blown away by the wind, some are carried away as burrs on the coats of animals, some are even eaten along with the fruit and survive the gastric tract of the animal involved.

Humans have been manipulating the genetic material of plants for centuries through selective breeding. Sometimes this has been little more than an emphasis of the processes of natural selection, but usually more than that is involved, with careful selection of pollen and of recipient fruit. In this way, varieties have been developed that are more attractive to humans in terms of food value, crop yield, flavour and disease resistance than were the natural species. Maintenance of the stock line, however, generally requires continued human intervention. If a new variety is simply planted and allowed to reproduce, it is most likely to receive pollen from other more common varieties so that it gradually reverts genetically. The variety can be maintained only by continuous vigilance, which generally means the use of new seed rather than what is produced naturally from the plant.

This process can be taken a step further by using pollen from one species on the flowers of another related species, producing a hybrid when the seed is grown. Many hybrids, however, will not breed true, either producing a mix of varieties or perhaps infertile seed. The hybridisation process must therefore be repeated with the two original species to produce seed for each new planting. Despite all this trouble, hybrids often have such desirable properties in terms of crop yield or crop quality that it is all worth while.

These plant-breeding methods have now been greatly extended by the use of what has come to be known as genetic engineering. The result is the same as hybridisation, but it can be planned in molecular detail and the opportunities it offers are much wider. As we noted in Chapter 11, the genetic material of all living things is derived from the same basic building blocks—the amino acids linked together to form the huge molecule of DNA. The DNA molecule has the form of a double helix which unwinds and copies itself in the process of cell division, the arrangement of links along the DNA chain carrying all the genetic information about the life form to which it belongs. If one does anything radical to a DNA molecule, such as cutting a large section out of it for example, then it will not carry enough information to produce a coherent living structure. It is possible, however, to modify a tiny length of the chain by adding part of a chain from some other species, and thus to produce a new individual with a tiny modification included in it. An example might be the modification of the poisonous repellant produced in the plant leaves—something not essential to the life of the plant but having an important effect in repelling predators. Either we might make the taste repellant to a new species of insect, or we might inactivate the repellant-producing gene so that the leaves are palatable to stock. In this way it is possible to produce genetic material whose cells can be multiplied to produce plants with some sort of desirable feature that would take years or even be impossible to introduce by selective breeding. The new plant will breed true to form provided that it receives pollen from another similarly modified plant, so that in this way a new variety can be produced.

It is important to realise that, although it is possible to transfer a tiny link of genetic material from one plant to another, from one animal to another, or even between a plant and an animal, it is virtually impossible to make any vast changes in this way—a radical change will almost certainly lead to a non-viable cell. There is, however, the demonstrated ability to slightly alter the properties of plants, in much the same way as one might do in selective plant breeding, to achieve beneficial results.

Plants as Energy

As we have seen, plants are living organisms that take simple molecules such as carbon dioxide and water and, with the help of energy provided by sunlight, build these up into much more complex sugar molecules. Subsequent biochemical processes in the plant then convert these sugars into starches, cellulose, and other complex materials necessary to the life and physical structure of the plant. It is not surprising that other life-forms have developed to take advantage of all this hard work, and particularly of the valuable energy stored in the structure of plants. These life-forms may be animals such as caterpillars or kangaroos, or microscopic bacteria either living in a cooperative way with the plant or waiting to devour its substance when it dies.

Plant energy is stored as complex molecules consisting mostly of carbon and hydrogen atoms with a small number of oxygen atoms. These molecules are valuable for many reasons. We have already mentioned their role as food sources for animals, and we return to this in the next chapter, but they are important in two other ways as well. The first is simply because of the stored energy, which can be recovered and used for other purposes by burning, which simply means converting the complex molecules back to the carbon dioxide and water from which they were made. In doing this in the most efficient manner possible, and with the most suitable plants, we might recover perhaps 10 to 20 percent of the energy absorbed by the plant from the sunlight. This important source of energy is likely to be used increasingly in the future with specially grown crops. For convenience, the plant material would probably be converted first to alcohol and then burnt at high temperature in power stations or motor vehicles.

More traditionally, of course, we burn fossil fuels such as brown coal or black coal, or the basins of crude oil, which are the remains of huge beds of vegetation that flourished about 300 million years ago. The heat energy available from oil is particularly convenient and quite inexpensive, but once the present oil supplies are gone there will be no more. Coal is still more plentiful and even cheaper, but again the reserves are not inexhaustible. Again of importance is the fact that converting these reserves back to energy necessarily consumes oxygen and releases carbon dioxide into the atmosphere. The consumption of oxygen is not a problem because the supply is so immense, but the release of carbon dioxide has already increased its concentration in the atmosphere by nearly 50 percent in the past 200 years, and there are growing concerns, which we discussed in Chapter 7, about the effect of this on the climate of the Earth.

The second way in which the stored chemical energy of plants can be used is also important. It comes from the fact that plants manufacture chemicals, not just for energy but also to repel predators and to perform a host of similar subtle tasks. The variety of these chemicals manufactured by different plant species is immense and, because humans are animals, many of them have an effect upon us. The effects can range from an appreciation of the perfume of a flower, through the disinfectant action of various aromatic plant oils, to the powerful medical action of drugs such as aspirin, quinine and hosts of more subtle plant derivatives. Considering their origin, it is not surprising that many of these, such as nicotine, mescalin, opium and curare are harmful or even deadly, but the number of medically useful plant derivatives is immense. A major task is simply to evaluate their effects on the human body in disease or health.

Plants, we must remember, have been on the Earth much longer than animals, and incomparably longer than human beings. Whether we realise it of not, they form the very basis of our contemporary life and we must both cherish them and take the good things they have to offer.

13

The World of Animals

"All animals are equal, but some animals are more equal than others.' George Orwell "Animal Farm"

In contrast to plants, which can live and grow on an input of simple molecules such as carbon dioxide and water together with light energy, animals require more complex and specific food and do not require light. The basic ingredients of animal food are large molecules containing carbon, hydrogen, oxygen and some nitrogen, and it is no accident that these are just the sorts of molecules manufactured by plants—sugars, starches and simple proteins. Animals have digestive systems to deal with these molecules and to break them into smaller units which serve as basic nutrients for animal cells. In the process of breaking down these large molecules, the animal system consumes oxygen, gives off carbon dioxide, and releases some energy as heat.

Since humans are animals, we are rather more interested in ourselves and other animals than we are in plants, but it is a reasonable comment to say that animals are more complex than plants so that there is more about them to study and understand. Let us look at some of these aspects of the world of animals.

The History of Animals

The early history of animals is contained in the fossil record, which can now be given dates with a good deal of certainty based on radioactive decay measurements. The study of fossils (paleontology) is a major part of both earth science and biology. Animal remains are a good deal more plentiful as fossils than are plants because they usually have boney shells or skeletons that are well preserved. The earliest animal fossils have been dated at about 600 million years ago, though there were certainly simpler animal-like organisms at much earlier times. These animals lived in seas or shallow lakes, and even at that time there was great variety amongst them—sponges, corals, snails, clams, crustaceans and many others. Some of these early animals look quite like similar animals today, but many of them have no modern counterparts.

It is convenient to divide animals into those with backbones (vertebrates) and those without (invertebrates). The earliest animals were all invertebrates, but about 500 million years ago the record shows the occurrence of cartilage fishes— ancient relatives of modern sharks—then boney fishes. About 350 million years ago amphibians began to colonise the shores, and reptiles developed about 300 million years ago. Birds have a history going back about 100 million years, while mammals have a slightly shorter record. Modern humans—*homo sapiens*—have been around for only about 200,000 years, though hominids of earlier types evolved as long as 4 million years ago. The rise of numbers of bird and mammal species occurs at the same time as a great decrease in the number of reptiles, particularly large reptiles such as dinosaurs which dominated the world for about 150 million years, a most impressive record of evolutionary success. We cannot yet be sure what ended the rule of the reptiles and allowed the development of mammals, but the evidence suggests that it was a great climate change, perhaps brought about by the impact of a large meteorite or a prolonged meteor shower.

Evolution of animal species by natural selection is still occurring, though the process is so slow that we are not aware of it. Indeed many species are so well adapted to their environments that almost all variations are unfavourable and their form has remained stable for millions of years. If, however, the environment changes for some physical reason, then there can be a rush of evolutionary change in the ensuing million years or so as species adapt. Evolution tends to run in stops and starts like this, rather than as a continuous process.

In the course of the long history of the Earth, many species have adapted in this way to new environments, but many have failed and have become extinct. A few have been so well adapted to a rather protected environment such as the deep ocean that their modern forms are nearly the same as those of a hundred million years ago. It is always a loss to the world as a whole when a species becomes extinct, for the unique pattern of information coded in its DNA is lost with it. Modern humans have accelerated this process of extinction by changing environments so rapidly that particular life forms have been unable to adapt. Even communities such as Australian Aboriginals contributed to this process when they migrated to Australia. Their hunting hastened the extinction of the giant marsupials, and their "fire management" techniques changed the plant environment so that some species thrived and some were destroyed. The introduction of European technology has vastly accelerated this process by destroying forest environments, and increasing pollution from the human species threatens many forms of life in rivers and oceans as well as on land. We owe it to our children to preserve as much biodiversity on the planet as we can.

The Animal Body

Animals are such complex organisms that it is useful to consider them in parts, just as we discussed plants in terms of leaves, stems and roots. The level of subdivision depends on our purpose, and here it will be enough to consider just a few of the major interacting systems that make up a typical animal body.

The structure of animal bodies falls into two main types, as we noted before—vertebrates with backbones and invertebrates without. In either case, with a few exceptions such as caterpillars, the animal requires some hard structures to support its softer tissues and to allow it to move, and the solutions evolved are either to have the hard tissue on the outside, as in insects, crustaceans and shellfish, or to have an internal skeleton as in reptiles and mammals. A hard shell has obvious advantages by way of affording protection from predators, but must ordinarily be shed several times during the growth process. Indeed in most insects advantage is taken of this necessity to allow considerable changes in external form, as when a caterpillar changes into a moth.

The most important internal systems are those that keep the animal alive—a system to take in and digest food, a system to take in the oxygen necessary for cell growth and to get rid of the waste carbon dioxide, and a system to carry nutrients to all parts of the body. We usually call these the digestive system, the respiratory system and the circulatory system respectively, and all are too familiar to require much detailed discussion. What is useful, however, is to look at the ways in which they differ between rather different classes of animals.

The digestive system is basically a chemical processing line which takes in complex animal or vegetable matter and breaks it down into sugars, proteins and other simpler but still complex molecules that are needed to feed the cells of the body. Animals that eat only vegetable matter need to eat a good deal because its energy content is rather low and it contains a great deal of matter that is not very useful—cows and horses eat nearly all the time. Meat-eating animals, on the other hand, take in food that is rich in energy and proteins so that they can take in a single meal and digest it at leisure. Each requires a rather different type of digestive system to cope with the different nature of the food. Animals such as humans can eat both plant and animal material,

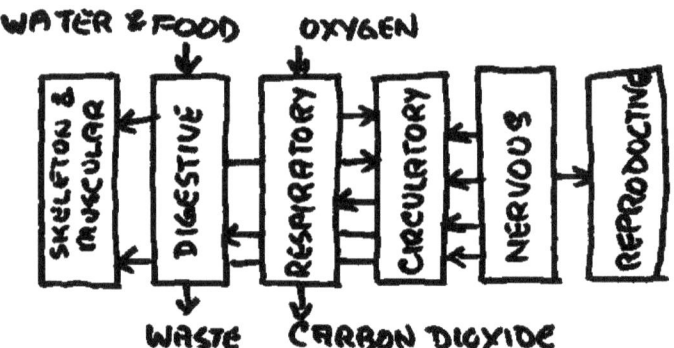

It is convenient to analyse the functioning of an animal in terms of systems that are moderately independent, though all are in reality interconnected.

and the same was true of our ancestor hominids and still applies to our cousins the primates. Insects are similarly specialised to eat either plants or smaller animals.

The respiratory system differs between small animals such as insects and larger animals such as mammals. In insects, all that is needed is a system of tubes running into the body which allow oxygen to enter and carbon dioxide to escape. There is no need for active breathing. Larger animals, however, need muscles to pump air into and out of the lungs where it supplies oxygen and takes away carbon dioxide. To circulate this needed oxygen to the cells of the whole body, a circulation system is needed, pumped by the heart. The blood itself is red because it requires a special chemical—haemoglobin—to carry the oxygen, which does not dissolve much in plain water. This circulatory system also serves to carry nutrient from the gut to all the cells and to return waste products to various central organs for excretion.

Some animals have evolved systems to keep their body temperatures approximately constant, a device that requires considerable energy but that allows the animal to be equally active in all kinds of weather. Others, such as reptiles and insects, allow their body temperature to follow that of the environment, with a great saving in energy but with the consequence that all their body processes, including movement, become very slow in cold weather.

Animals living in water have similar problems to those living on land and have evolved similar solutions. All digestive systems are very much the same in general design, and the gills of a fish are very similar to the lungs of a mammal. Most marine animals simply let their body temperature follow that of the water, which does not in any case change too much. A few such as whales, seals and dolphins, however, are more closely related to mammals living on land, and are warm blooded.

The fact that most animals can move independently also differentiates them from plant life, and to do this they require muscles. These are collections of cells that are able to change their shape from long thin fibres to shorter fatter fibres under some form of chemical or electrical stimulation. The ends of the muscle fibres are tethered to solid bone structures or connected to long tough cords called tendons, the other ends of which are tethered to distant bones. By an appropriate system of hinged joints (as in the elbow) or ball-and-socket joints (as in the hip or shoulder) the whole animal skeleton can be made to move in a coordinated way.

The Energy of Humans

It is interesting to have some idea of the energy consumed by humans in everyday life and of the physical power that we can command. An ordinary adult diet has an energy content of about 10,000 kJ/day. (This used to be called 2500 Calories per day. 1 Calorie, with a big C, is equal to 4.2 kilojoules and is also equal to 1000 calories, with a small c. A Calorie should really be called a kilocalorie, but this is getting too complicated!) This energy input works out to about 100 J/s or 100 W. This energy is mostly used to run our internal organs for breathing, blood pumping, etc., and appears as waste heat from the body. If we eat much more, then the extra energy input is stored as fat. Air conditioners in offices and theatres have to cope with this heat from their human inhabitants, plus any extra heat generated by machines.

Of equal interest is the muscular power that humans can develop. One of our most strenuous exercises is running up stairs. Suppose a person weighs 50 kg and can run up stairs fast enough to increase his or her height above ground by 1 metre per second. The power involved is the force (mass times the gravitational acceleration, which is $10\,\mathrm{m\,s^{-2}}$) multiplied by the speed against the direction of this force, so the answer is 500 watts. A fit adult, weighing more like 100 kg, can probably run upstairs at about the same rate and so produce about 1 kilowatt of power. In both cases the maximum power produced is about 10 watts per kilogram of body weight.

While these figures are quite impressive (in the old system, 1 horsepower was about 746 watts), a simple trial shows that we cannot keep this power output up for more than half a minute or so. Less energetic exercise, such as walking or running along a level surface, does not do any external work and the extra energy used is just because of the inefficiency of our muscles. Humans are actually not very efficient producers of mechanical power. Four legged animals such as dogs or horses do several times better per kilogram of body mass, and birds and insects better still.

Nerves and the Brain

This brings us to the subject of nerves, the brain and consciousness—a subject that could easily fill a whole library of books! Nerves are comparatively simple, because they can be studied experimentally, and even the brain is now revealing some of the secrets of its operation, but discussions of consciousness are still extremely complex.

Nerves are special cells with very long fibres connecting them to other cells. When a nerve cell is stimulated by some sort of energy input, it sends an electrical pulse along its outgoing fibre. Where the fibre connects to another cell it releases a tiny amount of chemical messenger material which in turn either stimulates this cell or inhibits it from doing anything. Because each nerve cells receives inputs from many other nerve cells, the whole connected network responds differently according to what the stimulus is and where it is received. Some nerve cells have very clear message paths, such as those connecting the eyes to the brain, but when we enter the brain itself the interconnections are so complex and extensive often that whole large areas of the brain respond to each stimulus.

It used to be thought that the brain behaved like a telephone exchange or like a computer, and that it had a built-in program learnt in childhood that determined how it responded. We now know that it is not as simple as that. We understand a great deal about the operation of the brain, but the details of the mechanisms of memory and response are still beyond us. On the other hand, we understand more about messages passed from the brain in the other direction. These messages consist first of electrical pulses sent along long nerve fibres and relayed by chemical messenger substances at nerve junctions until they reach their destination. Here the final chemical messengers generally cause a muscle fibre to contract, beginning the motion of a limb or initiating some internal process. Many processes such as the beating of the heart, however, are controlled by nerve pathways and electrical generators in the organ itself and do not require active control by the brain.

The extent to which the control of muscle movement has developed in animals is the result of a complex series of conscious and automatic control processes. The animal makes a general decision to do something, such as walk or fly, but details of the process are taken care of by automatic controls that are either built into the nervous system or else learnt by the animal when young.

Consciousness

If we think carefully about it, we can see that animals have a nearly continuous range of responses to stimuli about them. The very simplest animals such as shellfish simply close up if disturbed in what is clearly a reflex action rather like a heartbeat. Other animals, such as fish or mice, react in much more complex ways to their environment and, though they do not seem to think ahead, they are able to profit by experience. At the top of this particular tree comes humanity, whose individuals can not only think ahead and learn by experience, but can also communicate in complex language and worry about their place in the universe.

The whole idea of conscious mind raises many profound problems, and we shall not try to answer any of them here. Rather, let us look at a few simpler issues simply to get the flavour of the field. One is language. We know that human language expresses many of our ideas very well, but falls short when we try to say precisely what we mean in other areas. Do other animals have languages? Certainly chimpanzees come quite close to human thoughts and actions and are able to communicate quite well with each other, but their vocal utterances seem to be quite limited. Whales and dolphins similarly are able to communicate with each other, but we know so much less about their life and behaviour that it is hard to assess the extent of the communication. At the other end of the scale, many birds have rich and complex songs that may vary after extensive migration journeys. They sing much of the time—but do other birds listen and gain much information? It is by no means obvious that birds with rich songs are more intellectually developed than those that simply chirp.

Even behaviour is difficult to analyse. Consider, for example, the social insects such as ants and bees with their complex and highly organised societies. Is all this behaviour simply "instinct" or is there more to it than that? Without answering this question, let us look at just one example—the way in which ants cooperate to keep their nest tidy, with eggs in one pile, stones in another, and so on. Is this evidence of some group intelligence, or of intelligence at the individual level? Actually here the answer is "no"—the whole complex activity can be accounted for by the application of two simple rules:

1. If you find something, pick it up

2. If you find another of the same things, put your one down.

with the rule "run around" following each other activity. These two rules will ultimately sort everything into neat heaps as observed, but will not do it very efficiently—ants will continually take items from partly built heaps and carry them off—but this is just what we see happening. And in the end, the job will be done.

In the higher animals, however, particularly those such as dogs that have developed as companions to humans and those such as chimpanzees that are our close biological relatives, it is hard to come to any conclusion other than that their consciousness is very much like ours. We are all part of the same biological world, and our histories and our destinies are inextricably linked.

14

Living with Nature

"Nature, to be commanded, must be obeyed." Francis Bacon (1561–1626)

It has been recognised for a long time that all parts of nature are interconnected in a complex way, but the science of ecology has really developed only over the past fifty years or so. We now know that the Earth's ecosystems are under severe threat from human activities and that we must take urgent steps to make our use of the planets resources sustainable in the long run. We return to this important matter later in the chapter; let us look first at the way in which an ecosystem that is not under threat functions.

The prime requirements for any living organism are food and shelter, and both plants and animals have found niches in the complex web of life and its physical surroundings in which they have been able to find what is necessary for their survival. These niches are of tremendous variety, from animals living in burrows dug in scorching deserts, through huge varieties of life in temperate rainforests and around coral reefs, to lichens clinging to rocks on the freezing shores of Antarctica. Let us look at just a few general principles and representative cases.

Shelter

In all parts of life, the weak shelter in the shadow of the strong, and nowhere is this better seen than in a tropical rainforest. Huge trees form a canopy of sheltering leaves to filter the sun and retard evaporation, their fibrous roots bind together the soil, and their dead leaves form a rich compost in which smaller plants can flourish. Monkeys and other animals live high in the tree canopy, birds build their nests in branches, other animals prowl the forest floor and live in burrows or hollow logs, and there is a ceaseless traffic of insects of all kinds at all levels.

The trees are the major partners in this ecosystem—remove them and the habitat of birds and animals is destroyed, the nutrient leaches from the soil which then washes away, and the whole ecosystem dies. This does not mean that trees are sacred, but they are vitally important and they must either be protected or else used with an intelligent regard for their relation to the rest of life.

This relation between soil and vegetation is a widespread one, not just confined to rainforests. Trees in flat open country may play a vital role in maintaining the level of groundwater. Without trees the groundwater level rises, bringing salt to the surface and killing the grass; without grass the soil is eroded by wind and water and the land becomes arid desert.

At a lower level, the other members of the ecosystem also provide shelter for smaller partners. Parasites attack the bark of trees and make holes for shelter, animal parasites live in the dense fur of small mammals, or even inside their bodies, fungi live among the roots of trees and plants, and microbes abound everywhere on and inside other living things. Sometimes this close relation is a welcome one—ants protect aphids and harvest their milk, bacteria excrete soluble nitrogen compounds that can be absorbed by plants, other moulds and bacteria break down the leaf litter and turn it into useful compost. But many of the close relations are harmful to one of the partners—borers weaken the trunks of trees, parasites suck the blood of animals, and harmful microbes cause disease and death. Life is a constant struggle against forces tending to destroy it.

Food

The other thing that a balanced ecosystem must do is to provide food for all its members, and here we can draw some general conclusions. Plants provide the basic food for everything else in the environment and, since their food energy concentration is not high, they must exist in abundance to support the animals that graze upon them. The total live weight of animals grazing upon a given plant population is typically less than about one tenth of the live weight of the vegetation.

Ideally we might imagine a balance in which there was just enough food to support these animals, but this is not the way it works. In a good season the food supply is plentiful and animal numbers rise rapidly, but then the next

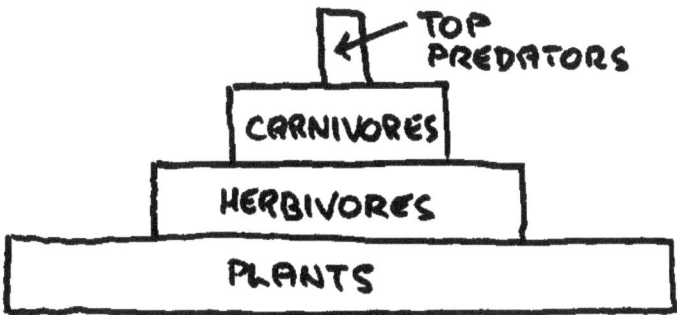

In a moderately stable ecosystem the populations at various levels in the food chain form an ordered progression. At each level there is a reduction of about a factor ten in total live weight compared with the next lower level.

year, or the year after, there may well be too many animals for the vegetation to support. The plants are eaten down to a small reserve and the animals simply die off. With a reduced animal population, the plant food supply recovers, and the cycle is ready to repeat itself. Such fluctuations in food supplies and populations are very common, particularly in ecosystems in which there is not a large variety of alternative food resources. Mice can be a particular problem in wheat-growing areas because their numbers can grow rapidly enough to be called plagues. Of course, because wheat is a crop cultivated by humans for our own use, drastic steps are taken to reduce the mouse population.

At the next level in the food chain we find the carnivorous animals that feed upon the herbivores. The same sort of principles apply—the weight of predators eating a given population can only be about one tenth of the weight of that population, and their numbers are subject to the same sort of fluctuations we found for herbivores. Sometimes the population of predators goes through fairly regular cycles of a few years duration simply as a result of the lag between plentiful food and population increase, but of course a severe environmental event can upset this completely.

At higher levels in the population we find animals that prey only on particular species of other animals, and for these a similar pyramid relation between the live weights of the populations must also occur–the top predators, such as

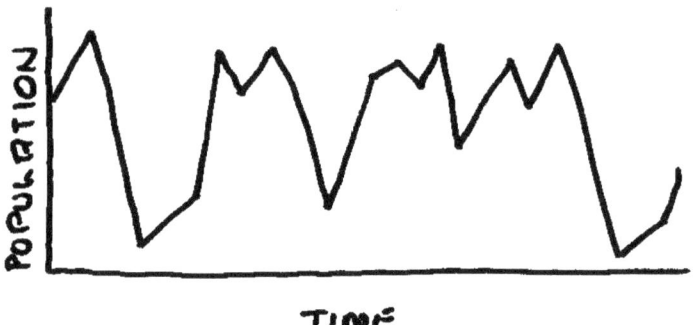

In ecosystems with relatively simple food chains there can be rapid and quasi-cyclic fluctuations in population numbers with periods of a few years.

lions or eagles, are a select aristocracy.

Managed Ecosystems

Humans, of course, have the ability to plan, and one of the earliest things they planned and changed purposefully was the ecosystem in which they lived. This was the beginning of agriculture and the beginning of modern civilisation and took place nearly 100,000 years ago, quite a long time after evolution of the species *homo sapiens*, which has been around for perhaps 200,000 years. Agriculture began with collecting seeds and planting them in places that were conveniently accessible and well watered. It extended to the capture and domestication of animals for food, and later to the selective breeding of both plants an animals to make them more suitable for human use. Settled agriculture then led to the development of villages and towns, and an assured food supply gave leisure for the development of arts and crafts.

We are not concerned here with the social history of these developments, but rather with the nature of the managed ecosystems that humans have created. How, for example, do they differ from natural ecosystems? The first thing we might notice is that a managed ecosystem requires continuing management—if a farm is abandoned then it quite quickly reverts to a quite different natural ecosystem, generally with many more species in abundance. Managed ecosystems tend to be monocultures, such as fields of wheat, or to have one dominant species, such as on sheep or cattle stations. Such an arrangement is clearly of great practical convenience, but usually brings with it a requirement for further technological intervention.

The problems with monocultures are basically two. The first is that repeated growth of a single species often exhausts particular nutrients from the soil. The age-old solution to this problem was crop rotation, a technique developed more than a thousand years ago. By carefully choosing the crops in the rotation, and perhaps leaving the fields fallow for one year in four, nutrients such as soluble nitrogen compounds can be returned to the soil in much the same way as if three or four contrasting species occupied the ecosystem at the same time. There is some move to return to this system as sustainable agricultural practices are developed, but the more usual solution is to maintain soil quality by adding fertilisers.

The second problem is that of predators and diseases, both of which can spread with great rapidity in a monoculture environment because of the availability of such plentiful supplies of suitable food. Crop rotation can help a little with this problem, particularly if plots are relatively small and are not rotated in the same cycle, but again the most common and effective solution is with the aid of insecticides and fungicides. In the case of animals, veterinary medicines play a similar role.

These practices are not "bad" in any scientific sense, and indeed they are usually essential if we are to maintain these managed ecosystems in a productive state. However attractive it might seem to return to "natural" agricultural methods from a philosophical point of view, it is simply not possible to produce enough food for the world's expanding population without managing the agricultural ecosystem in a very positive way. This does not mean, however, that we cannot do much better than at present by using our knowledge of natural ecosystems. Natural predators may offer a better solution to problems of pest control than do chemical sprays, judicious crop rotation or even co-planting may take the place of heavy fertiliser application, and the new techniques of genetic engineering can enable us to produce varieties of common crops that are resistant to particular diseases or predators.

The Problems of Overpopulation

Of all the problems of the world's ecosystems, the most serious is the huge expansion of human population in recent centuries. For tens of thousands of

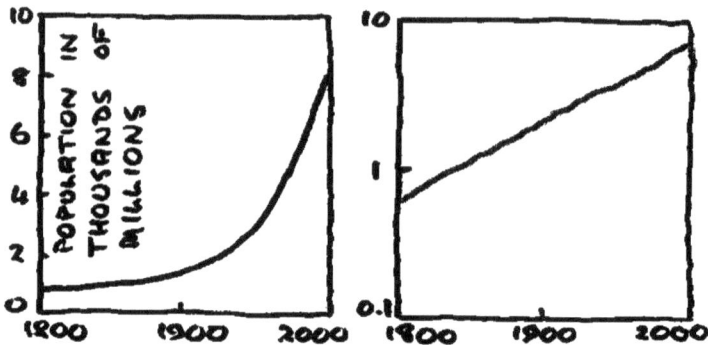

The growth of human population is devastatingly rapid. When plotted on a logarithmic scale, as at the right, the trend is more easily extrapolated. Unless the human birthrate is decreased very soon, there will be a catastrophic exhaustion of resources, just as is seen in other animal populations.

years humans lived "in harmony with nature", which sounds pleasant but is not so in reality. Before the development of modern science and technology, human life was quite short and, much more importantly, infant mortality was very high. This kept the human population in balance with nature in a very effective way.

The problems began first with agriculture, which allowed larger groups of humans to survive, and then with advances in drainage and water supply technologies which reduced the incidence of disease. Modern medicine, and particularly the invention of antibiotics, have now reduced infant mortality to such an extent that the human population is expanding with huge rapidity and almost inevitably polluting and destroying the ecosystem that supports it. The "harmony of nature" has been destroyed through the use of humanitarian weapons.

The only solution is, of course, to reduce the human birthrate to match the deathrate so that the population becomes stable. This tends to happen to some extent naturally, but the rate of decrease is far too slow to save the situation. Science has developed methods of birth control, but their use is opposed by some people on the basis of religious tenets developed more than a thousand years ago when humans were desert tribes struggling for survival in a harsh environment. Even with the most active measures, the population of the Earth is likely to double before it stabilises, and the increases will be greatest in those countries least able to support them. If there is any substantial shift of rainfall zones because of the greenhouse effect of extra carbon dioxide in the atmosphere, then this will make the problems much more severe.

It is a challenge for the social sciences to develop an effective approach to this world problem—to accelerate the reduction on birth rates, to find ways to distribute equitably the available food and other resources, and to avoid conflict. Science has provided most of the tools needed to do this job, but it is the task of sociologists, economists and politicians to use these tools for the benefit of the whole of humanity.

Water

Food and water are the prime requirements for animal life, and they are closely linked through the need of crops for water. At a more direct level, however, modern life depends critically upon having enough water of high purity for drinking, washing and running sewer systems. All this water must be somehow collected from rainfall, stored, and then distributed to individual houses. Ancient cities were built around places with reliable river water supplies, but it was not until the time of the Romans, a little over 2000 years ago, that civil engineering developed to the stage of bringing water large distances in stone aquaducts and reticulating it to individual houses in lead pipes. Some of these mighty engineering works can still be seen in Italy and the south of France.

In the ancient world the population was so small that these water supply projects had very little impact upon the environment, but all that has now changed. Many of the world's rivers have been dammed to store water for irrigation or domestic use, and this has changed their flow so dramatically that ecosystems downstream of the dam are severely affected. With ever growing population, still more dams must be built, and they must be planned carefully from both engineering and ecological points of view. The engineering must ensure that the dam is strong, that it leaves an adequate flow downstream of the dam, and that it does not quickly fill with silt. The ecological planning must ensure that the water use does not destroy the soil through salting, and that the changed river flow pattern does not lead to too much destruction of native ecosystems.

It would be nice, of course, to be able to leave all the world's rivers and ecosystems in their natural states, but that is simply not possible without condemning more than half of the world's growing population to a precarious existence on the edge of famine. Closer to home, our cities simply would not function without adequate supplies of clean water. Of course, we can be much more careful than we are at present with water usage, and we can clean and re-use much effluent water for less critical purposes, but still water will constitute a problem as great as that of food in the future until our population has stopped growing.

Energy

We discussed many of the properties of energy in Chapter 3, and it is clear that energy is essential for the maintenance of all natural systems. Energy is even more directly important for the maintenance of civilisation. While small communities with stable populations can (or at least used to be able to) live delightfully on Pacific islands in a simple hunter-gatherer style, this is not possible for large communities in cities. There are some who advocate a return to "alternative" low-energy lifestyles away from cities, but such a luxury is possible only for small minorities who then rely upon the bulk of the population for essential services and manufactured goods.

The primary sources of energy to support civilised life are at present fossil fuels—coal and oil. These are inexpensive commodities and are very convenient because of their high density of stored chemical energy. Certainly extracting them from below the ground involves minor environmental damage, but that is a very small price to pay. Perhaps more important is the change to the environment that seems likely to come from increased carbon dioxide emissions, but it is too early to be definite about the effects, as we discussed in Chapter 7. The major concern is that we are consuming all the reserves of highest quality first, and their amount is limited. When the oil wells are exhausted—and this will certainly happen sometime in the next 100 years at our present rate of use—it will be necessary to recover oil from deposits of much lower grade, such as shale. This will raise the cost greatly and will increase environmental problems because of the larger areas that have to be mined. Much the same will happen with coal, though on a longer timescale.

If we look to human civilisation persisting for at least another thousand years, which is not very long, and hope that human population may stabilise at not much more than twice its present level, then we will need to find other sources of energy to run our civilisation. The long-term options seem to be two—to use some form of nuclear energy or to employ solar energy. Let us look briefly at these two alternatives.

Nuclear energy can be derived from the fission of uranium in nuclear reactors.

Natural uranium, which is reasonably plentiful in Australia, is a mixture of two isotopes with atomic weights 238 and 235, and it is the isotope 235 that provides the primary fission reaction. The natural uranium must therefore be processed to remove some of the 238 isotope and concentrate the 235 before it is used in reactors. A reactor simply produces heat from the nuclear fission of uranium, and this heat can be used to make steam which drives turbines to produce electricity. Some reactors, called breeder-reactors, are arranged so that they produce a new radioactive element plutonium from the 238 uranium isotope, and this plutonium can be separated and used as fuel for other reactors.

The whole process is now easy to control, but presents several problems. The first is that the enriched uranium or the plutonium could be used to make bombs. Such a threat is always present, particularly in the hands of terrorists, and succeeds because of the dramatic power of nuclear explosives. In reality, however, biological toxins placed in water supplies could be even more dangerous. The second problem is that of reactor safety—not from explosion but from possible failures leading to the release of radioactive material into the atmosphere as happened in the Chernobyl accident. Clearly reactors must be designed and operated safely, but chemical plants can present almost equal hazards. Finally we have the problem of disposal of small amounts of radioactive wastes that may remain hazardous for hundreds of years. Techniques that seem reliable have now been developed, such as dissolving the wastes in synthetic rock minerals which are then stored in carefully selected underground sites. Parenthetically one might remark that, after a hundred years or so the radioactive materials become harmless, while some other industrial contaminants such as arsenic remain hazardous for ever.

There is another approach to nuclear power that is still attracting a good deal of research, and that is power developed from a fusion reaction converting hydrogen into helium, as in the reaction that powers the sun. Here the starting and finishing materials are harmless, and a great deal of energy is released, but the problem is to make the reaction run. Instead of a simple grid of uranium fuel rods immersed in water, we need to heat hydrogen gas to immense temperatures in strong magnetic fields. It is likely that this technique will prove successful sometime within the next fifty years, but its economics have not yet been evaluated. It is inherently safe, at least from explosion or the risk of large radioactive contamination, but the whole reactor vessel will become slightly radioactive, so there will be storage problems after decommissioning.

Seen beside nuclear energy, power derived from the sun appears very clean and safe, and indeed this is true. The problem with solar power is simply that it is spread rather thinly over the whole surface of the Earth rather than being concentrated, so that large collecting structures will be necessary if it is to be a major source of power. This in turn means that there is a large cost involved in building the necessary collectors and associated equipment. To put this in perspective, a city the size of Sydney would need to collect and use all the sunlight falling on an area about 10 kilometres square to provide its electricity requirements. This is perhaps not a huge area, but it is comparable with the area of all the roofs in the city, and it would need to be a maze of mirrors, collectors and steam pipes. On top of this, some means would need to be found to store enough electrical energy to last the city through the night. This could perhaps be done with two large lakes, one much higher than the other. Water could be pumped to the higher lake during daytime and run back through hydroelectric generators during the night. If the height difference between the lakes were 100 metres, and each was 10 metres deep, then they would each need to be about 5 kilometres square to hold the necessary water with no reserves.

These figures are not meant to be reliable, but are intended to show the scale of structures that would be necessary and to emphasise that solar power will not be cheap. This is, in fact, its major problem—the energy is free, but harnessing it costs a great deal of money in construction and maintenance. Nevertheless, solar power is likely to provide a good fraction of our energy needs in the future.

We could make similar estimates for other ways of harnessing solar energy. The most common at present is the use of hydroelectric power, which is convenient and pollution-free and uses readily available technology. We might even take the view that it is completely sound from an environmental point of view, but this overlooks the fact that rivers in the high mountains must be dammed to provide the necessary water storage, and this has undesirable effects. It reduces the natural water flow to lower reaches of the river, with consequent necessary ecological adjustments, and it has an even more catastrophic effect upon the region that is actually flooded for the dam. These are, however, the sorts of compromises with which we are necessarily faced when finding ways to ensure a good standard of living for a rapidly increasing population. The decision is one to which scientific understanding can make an important contribution, but ultimately it is a social and political problem.

Other possibilities include using wind power collected by huge windmills on mountain ridges. Ocean based systems using tidal energy from the drag of the moon, or thermal systems exploiting the temperature difference between surface water and that deep in the ocean are also possible and experimental systems have been built. The environmental problems are again not negligible, however, and relate to visual and noise pollution and to the destruction of the natural environment at the site of the power station. Generally speaking, the magnitude of these effects is probably similar to those in the case of hydroelectric schemes. At the other end of the engineering scale, growing land plants or algae in ponds and harvesting their biomass for use as fuel after drying in the sun is quite attractive. Although the combustion of the fuel releases carbon dioxide, it is only in the amount that is removed from the atmosphere to create the biomass, so that the nett effect is zero. Each possibility is attractive in some particular situations, but it is not clear that any one holds the key to our future energy needs.

15

Materials and Structures

"Here's metal more attractive." Shakespeare "Hamlet"

Civilisation relies upon tools and structures and upon the materials from which they are made. Today we have an immense range of materials to choose from, and it is important to see how the design of structures depends upon the materials used.

When building structures, the things of greatest importance are the weight and strength of the materials to be used. Actually rather than weight we should think of density, which is mass per unit volume. Density is generally expressed as kilograms per cubic metre, and for reference we note that water has a density of 1000 kilograms (or 1 tonne) per cubic metre. It is sometimes useful to think of relative density, which tells us how much heavier a material is than the same volume of water. Rock, for example, has a relative density of about 3, while iron has a relative density of about 7.5. Aluminium alloys have densities around 2.7 times that of water. Wood, which generally floats on water, has a relative density between about 0.5 and 1, depending on the species of tree from which it comes. All these numbers are the equivalent of tonnes per cubic metre.

The reason that the weight of structural materials is so important is that the structure must first support itself before we add any people or other useful things. If we use heavy materials then the structure itself must be very solid. The ideal structural materials are therefore very light and very strong.

Compressive Strength and Design

There are several sorts of strength that must be considered in structural materials. The first is crushing or compressive strength and the second is stretching or

Materials and Structures 131

tensile strength. Rock or concrete both have high crushing strength, so that we can pile up rocks or concrete blocks without the risk that the structure will fail by crumbling, but have rather low tensile strength. This is even more true when we consider the way in which we assemble rocks or bricks by cementing them together with mortar. The mortar is quite strong in compression, and even if it fails the layer is thin and the rocks just move together. Timber beams, on the other hand, have good stretching strength as well as considerable crushing strength, and the same is true of steel and other metals. Equally importantly, we can join timber or metal beams together using bolts or other fixings that preserve this strength. As a final structural element we might consider ropes or chains. These may have good tensile strength but have no compressive strength at all—they simply fold up.

If we are restricted to building a structure from rock or plain concrete, then it is important that it be designed so that all the forces are compressive. The Egyptians built huge stone pyramids that were stable and strong, but when they wanted to make a doorway they relied upon the strength of a large single rock, much in the way that the ancient Britons built Stonehenge. This placed severe limits on what sort of structures were possible, and the pyramids were largely solid rock. The Egyptians did, of course, have other buildings that enclosed large spaces, but these were built with stone walls and the large distances spanned with solid timber beams cut from tree trunks.

The great innovation in stone design was made by the Romans who invented the stone arch. By making each stone of the arch in the form of a triangle with the point removed, the whole structure was pressed together under its own weight and no tensile forces were involved. A well-designed stone arch requires no mortar and will last for thousands of years, as is shown by the huge multilevel Roman aqueducts that still stand in southern France. Exactly the same design principles were used in stone bridges as little as a hundred years ago, and would still be used today except for the fact that cheaper and more effective designs using other materials have been developed.

As well as using stone arches in engineering works, the Romans also invented the stone dome, which is made in the same way, and built ceremonial buildings and temples with very large domed roofs. Again, some of these have stood for nearly 2000 years, and later architects have used the same design principles to build stone domes on religious and ceremonial buildings around the world.

The pinnacle of design using stones in compression was reached in the mediaeval cathedrals of Europe, which are immense stone structures with multi-arched stone ceiling vaults supported on thin stone pillars and with great coloured glass windows supported in thin stone tracery. While a few of the highest and

The Roman arch was one of the greatest structural innovations of engineers in the ancient world. Stone blocks are cut so that all the material is in a state of compression, in which it is very strong. Arches were cascaded to form massive structures such as aqueducts for the water supplies of cities.

most ambitious of these cathedrals collapsed while they were being built, most have survived and demonstrate the beauty and efficiency of design that can be achieved by masons with architectural sensitivity and a long tradition of technical skill. It is only the ravages of war, the insidious corrosion of industrial pollution, and the effect of changing water levels on the foundations that endanger these buildings which have stood for sometimes more than 800 years.

Tensile Strength and Design

Design of large structures in metal became possible rather more than a hundred years ago when large quantities of relatively cheap cast iron became available from the new blast furnaces. While cast iron is somewhat brittle, this is not of great importance in building stationary structures, though today we would use the much tougher steel.

Although iron is much heavier than wood, it is very much stronger, and a cast iron beam supports itself over a greater span than an equivalent wooden beam. Even more importantly, all iron beams can be made exactly the same and, apart from surface rusting which can be controlled by painting, they last nearly for ever. Iron can also be cast into plates and other fittings, and the

Materials and Structures

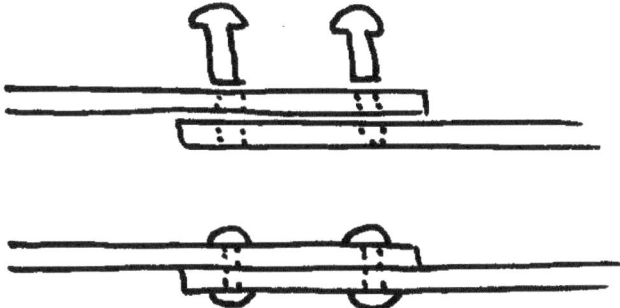

Girders of cast iron or steel can be readily fixed together by the use of rivets, which are inserted hot and hammered to produce an expanded end and fix them in place.

whole structure can be fastened together with iron rivets. These are simply short lengths of iron rod, with a thick head on one end, which are heated to make them soft, passed through holes in the structures to be joined, and then hammered to create a thick head on the other end.

When building a structure with beams, the main sort of distortion to be resisted is bending under the weight of the beam or of the load it supports. For a beam to be resistant to bending it needs to be wide in the direction in which bending might take place—it is much easier to bend a strip of metal along its thin direction than along the direction in which it is thick. We could simply make the beam solid and wide, but this would also make it heavy. Engineers developed a set of excellent design solutions to this problem. In the first solution, the beam cross-section is made like the letter I, so that the result is called an I-beam. Most beams used in building construction today have this form, although sometimes U-beams or even hollow-box beams are used for special purposes.

The second solution is to go even further and cut out most of the upright part of the I of the beam. Actually the process is not generally carried out in this way, except in spars for aircraft, but instead a complex beam or girder is made up by riveting or welding together two long bars and a number of oblique connecting struts. This sort of framework girder or truss was widely used in railway bridges up to about fifty years ago, and similar timber structures are now used in the roof trusses of houses. It is very strong and at the same time

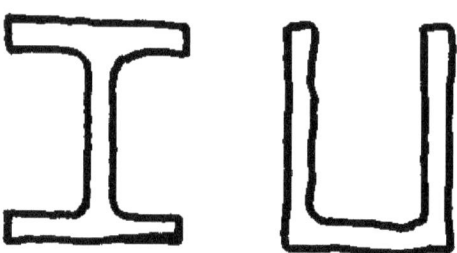

Special shapes are used to produce metal beams that are very stiff but not too heavy. Such beam sections are widely used in the construction industry.

very light because the individual beams from which it is made do not have to be very thick. In designing girders of this type, the fundamental principle is that a triangle cannot be pushed out of shape, whereas a rectangle can be. The girder is therefore essentially made from interconnected triangles. The pinnacle of such framework design was reached in the nearly spherical geodesic domes of the American engineer Buckminster Fuller, now widely used in exhibition buildings and sports centres, as well as in domes for radar installations.

When building a structure in metal the engineer designing it has a large choice of materials, for there are very many metals and alloys available commercially. Many things come into the choice, such as cost, durability and weight. Steel is generally the material of choice for large structures because it is strong, cheap and easily worked. Steel components can also be readily joined together by welding, which is essentially a process of melting the two joining surfaces together, using a hot gas flame or an electric arc. Stainless steel would have the additional advantage of resistance to corrosion, but it is quite expensive and so is used only for special purposes and for decorative architecture, like the huge flagpole on top of Parliament House in Canberra.

Steel is also generally used for ships and railway rolling stock, though fast suburban trains and small boats are now often made from aluminium alloys because the advantages of weight savings outweigh the extra cost. For aircraft, of course, weight is a prime consideration, and alloys of aluminium, or even

Framework girders or trusses are used to span large distances, such as involved in bridges or in roof construction. The material used may be either metal or timber.

magnesium which is still lighter, are used.

For modern aircraft, and some modern cars and small boats, new composite materials have been developed to replace aluminium alloys in panels and cladding skins. The most common in the case of boats is fibre-glass, which consists essentially of several layers of cloth, woven from glass fibre, impregnated and stuck together with a special plastic polymer. In the case of aircraft parts, special fibres made from glassy carbon are used instead of ordinary glass. These fibre composites are rather lighter than aluminium and very much stronger in some applications.

Another very important composite structural material is reinforced concrete, in which steel rods are cast into the concrete to give it tensile strength. In a further development, called pre-stressed concrete, these steel rods are pulled tight by tensioning nuts so that the whole concrete beam is in a state of compression. When the beam is later put under tension in a structure such as a bridge, this still does not exceed the original tension so that the concrete itself remains strong. Using these techniques, very elegant concrete structures can be built that combine the design freedom of steel with the aesthetic properties of smooth concrete. Modern bridges and buildings make extensive use of these techniques.

Structures and Stability

Stability is an important consideration in the design of any structure. For buildings that can be anchored in the ground, the main cause for concern is that the building might be tipped over by extremely high winds or by the shaking of an earthquake. For ships, and cars we add the requirement of stability when going round corners at speed, while for aircraft all possible motions need to be considered.

If we think just of structures such as buildings or towers that are stationary, the forces acting on the structure are its weight and the supporting forces exerted by the ground. To test for stability we have to consider the possibility that the structure experiences a tilt because of a gust of wind or other disturbance, and we want the design to be such that the structure will return to its normal position rather than tipping over. We can easily experiment with a chair of the type that has only four legs. If the chair is not very heavy, then the main weight involved is that of our body, and this can be considered as concentrated at a point called our centre of gravity which is somewhere about waist level. If we tip the chair only a little then it returns to its original position, but if the tilt is great enough that the position of the centre of gravity is outside the square formed by the four legs, then the chair will fall over. If we stand on the seat of the chair then it is much easier to overbalance than when sitting down, and this is a simple consequence of the fact that it then takes very much less tilt to move our centre of gravity outside the support are of the four chair legs.

This little experiment should convince us that, for a stable structure, we should arrange that the supporting base should be wide in the direction in which tipping might occur, and also arrange that the centre of gravity of the whole structure should be as low as possible and located fairly squarely over the supporting base. This second requirement can be met if we taper the structure so that it is narrower at the top than near the base, and at the same time arrange that anything particularly heavy is placed as close to the base as possible. If you are used to climbing ladders you will see that this agrees with experience.

It is not immediately clear that the designers of modern buildings take much notice of these stability considerations, since many buildings are quite high and essentially the same width at the top as at the bottom. Such buildings rely for stability upon the fact that they are firmly anchored in rock at their base, rather like a tooth, so that stability is not a problem, only strength. There are, however, many buildings in which stability has clearly been important in the design. One of the most obvious is the Eiffel Tower in Paris, which is a huge and elegant iron girder structure with a tapering shape and four widely spread

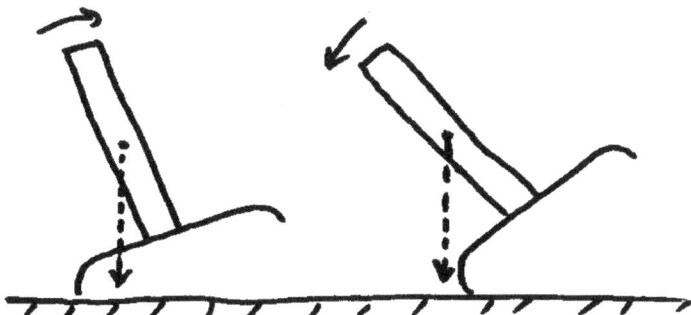

A structure will return to its original position when tilted if its centre of gravity is still above the outline of its base supports. If the centre of gravity is outside this area then the structure will overbalance.

supporting legs.

In a much more domestic setting, these same principles apply to the design of chairs and tables, standard lamps and refrigerators, all of which need to be stable under the disturbances and jolts they receive in household use. Vases of flowers present a particular problem because the vase is generally designed with aesthetic rather than structural considerations in mind, yet it must accommodate a rather wide variety of floral arrangements. Something heavy placed in the bottom of the vase is a great assistance to stability, as is the water in the vase, but at the same time it is usually necessary to balance the flowers themselves so that their centre of gravity is above the rather small base of the vase. An asymmetric arrangement or one with really heavy flowers generally requires a heavy vase with a rather wide base.

Biological Structures

Although the structures we have discussed above are inanimate things like bridges and buildings, exactly the same principles apply to the evolution of biological structures. Here it is not a case of deliberate design, of course, but rather the working of forces of natural selection to eliminate structures that were weak or unstable. The structures and environments of plants and animals are so varied that we cannot hope to discuss more than a few general principles governing their evolution and a few specific examples, but the same principles apply in all cases.

In the biological world, and particularly in the animal world, we do not have the luxury of a wide variety of structural materials from which to choose. First there is soft tissue, which has almost no structural strength, then ligaments, which have tensile strength but no compressive strength. There are only two real structural materials in the animal kingdom—the bone of vertebrate skeletons and the chitin of invertebrate cuticle. From one of these two materials, with the stabilising help of ligaments, all animal structures are built.

Because of this commonality of structural material, it is sensible to think first about the scaling of animals in size. At first glance we might think that this is simple—to make a bigger animal you simply double all the dimensions of a smaller animal, for example—but a little thought shows that this will not work. Suppose we were to do this, then the volume and mass of the animal would increase by a factor 8. The strength of bones, however, depends upon their cross section, which increases only by a factor 4, and indeed things are worse than this because the bones are longer. We need not worry too much about the details, but it is clear that a large animal needs to have bones that are proportionally much thicker than in a small animal, while at the same time its legs should be shorter if its skeleton is to have the same structural strength.

We can see this principle at work if we compare the size and shape of the legs of a mouse with those of an elephant. It is even more graphic if we can visit a museum and look at the skeletons of some of these animals. The same sort of thing applies to the external skeletons of insects. A tiny spider or a mosquito has extremely thin and delicate legs; but spiders of larger species are progressively more squat and solid, and the delicate structure of a mosquito becomes the more rugged structure of a large fly or a cicada.

If we look at the bones of any animal, for example in the meat chopped up by a butcher, we can see that the bones are not solid—some of them are hollow tubes, and all of them have a rather porous structure. There are several reasons for this, but one is related to the lattice girders and trusses that we discussed in connection with bridges—we can make a strong and light structure by having strong edges and a network of lighter members inside. Any reduction in weight that does not compromise strength and stability is clearly of advantage to an animal, so it is not surprising that bones developed in this way. This principle is carried furthest in the case of birds, as we might expect, where light weight is of immense advantage.

We could go further to look at the ways in which branches join onto trees and the ways sinews connect onto bones, at the position of the centre of gravity in relation to the legs and the way in which the animal moves, and at the problem of stability in walking on two rather than four legs, but at least this gives a general feel for the fact that the strength and stability of structures is as important in the biological world as it is in the physical world.

16

Tools and Machines

"Give us the tools, and we will finish the job." Winston Churchill

As the brains of animals became progressively more developed, some of them began to use objects in their environment to help them achieve particular purposes. The simplest such activities were simply the collection of suitable materials for nests or other shelters and this does not concern us here, but some birds, for example, use stones to help them open shellfish. By the time we come to advanced primates such as chimpanzees, the use of tools in this way is much more common—sticks are used to extract ants from nests or honey from hives. Other reasonably advanced animals such as dolphins or dogs have not developed these skills, presumably because of their lack of suitable hands for gripping. Thus while the use of tools is not an exclusively human characteristic, its development is quite recent and it is certainly something that humans have perfected.

We could think of a tool as being anything that extends the unaided capacity of a human. Generally we have in mind something physical such as a spanner, but something that helps us see more clearly or helps us to work out a complicated problem could also be called a tool. Bulk materials, such as washing detergents, are not generally regarded as tools, even though they help us to do things that we could not do as well unaided. A machine is a complicated tool, generally with some moving parts. Again we think of a sewing machine or a motor car as typical examples, and these have in common the fact that they contain a source of power, though this is not true of a machine such as a bicycle. There are no scientific definitions of these terms—they are simply ordinary English words—and it often makes sense to use other words instead such as "device" or "instrument" where this conveys our meaning more clearly.

There are so many different types of tools and machines that we cannot even hope to survey many of them. Instead it makes sense to look at a few general principles.

Effectiveness

Since a tool or a machine is used with the aim of extending what is possible for an unaided human, the first thing we need to ask of any tool or machine is how effective it is. Early tools were things such as spears, stone axes and knives for killing animals and cutting meat—humans are not very strong animals, and these extensions of their powers were extremely valuable. The effectiveness of these tools was put to practical test every day. The same was true of more sophisticated tools such as bows and arrows and blowpipes for poisoned darts.

The best way to assess effectiveness is to put it to a practical test. If a pair of spectacles is designed to improve our defective vision then we check whether we can read more easily when wearing them; if a telescope is designed for astronomical purposes then we use it to look in detail at planets, stars and galaxies; if a spanner is designed to help us remove tight nuts then we try it out. Usually the science behind tools is now well enough understood that we can calculate in advance just how they will work—the spectacles will be matched to the carefully measured defects in our vision, the telescope will have known magnification and light-gathering power, the length of the handle of the spanner will be calculated to give the desired turning moment on the nut, and so on.

Those who manufacture tools and machines are, of course, mainly interested in selling them, and it is not unusual for the advertisers to make exaggerated claims about their effectiveness. In some cases, as with medical products for example, governments regulate what can be said in advertisements; in other cases organisations such as the Australian Consumers' Association conduct tests of the effectiveness of many products and publish their findings; often the evaluation is left to the purchaser.

Efficiency

We introduced the "Plain English" term effectiveness above because it means something rather different from efficiency. When we say that something is efficient, we mean that not only is it effective in doing what it is supposed to do, but also that it does it with a minimal amount of waste. In many cases this judgment, like effectiveness, is rather subjective, although we might easily compare two machines that were supposed to achieve the same thing, for example two bicycles both intended to get you from place A to place B.

In some cases, however, we can actually make quantitative measurements of efficiency in terms of necessary energy input to the machine for a given task. Thus the efficiency of a car could be specified by giving the distance travelled on one litre of petrol—for some reason the published performance is specified as litres per hundred kilometres, which is just the inverse of the efficiency! However we might take the view that the object is to transport people, rather than just the car, so we might multiply this efficiency figure by the number of people transported to get a more useful number. Some consumer devices, such as washing machines, are now given an energy-efficiency rating in rather the same way.

Useful though these measures are, they are not quite what we mean by efficiency in a scientific sense. Some machines, however, convert energy from one kind to another—for example a power station might convert the thermal energy of coal into electrical energy—and then it is possible to be quite exact about what we mean by efficiency. It is simply the energy output divided by the energy input, and it is always less than unity, or equivalently less than 100 percent, and has no units attached to it. Where does the lost energy go, since we saw in Chapter 2 that energy is never actually destroyed?

Some machines are nearly ideally efficient. An electric motor, for example, may transform 95 percent or more of the electrical energy it uses into mechanical energy. In such machines the major cause of inefficiency is friction within the machine itself, and the energy lost in this way goes to heating up the machine in a non-useful way. Machines that take in their energy in the form of heat, however, such as coal-fired electricity generating stations or petrol-powered motor vehicles, are necessarily much less efficient. The reason is that, in order to work, these machines must take in heat at high temperature—that of the burning fuel—and give out heat at some lower temperature—that of the condenser. The maximum possible efficiency of such machines is related to these two temperatures—the further apart they are the better—but it is extremely hard to design a heat-machine that achieves much better than 50 percent efficiency. This is because we can't make the condenser temperature much less than ordinary air temperature—we would lose even more energy if we used a refrigerator for it—and there is a limit to the temperature that we can get from the burning fuel, set by available materials among other things.

We have already discussed some other examples of energy efficiency in Chapter 2, including the efficiency of biological systems, and we won't repeat this material here.

Machines

Although many simple machines such as egg-beaters and hand drills rely upon human power for their operation, the greatest advances came when other energy sources were used to drive machines. The earliest machines were probably water wheels, and these are interesting examples of reversible machines. Either we can use the energy supplied by water flowing from one level to a lower level to drive the wheel and supply mechanical energy for grinding corn or doing other useful work, or we can use the energy of a horse or other animal to turn the wheel and use this to move water from a lower to a higher level. Both forms were used in the ancient world. Windmills have similarly been used to extract useful mechanical motion from moving air, but it is only recently that turning propellers have been used to pull aircraft through the air.

Most machines have rotating wheels somewhere in their mechanism, and these are of very ancient origin. Wheels on chariots and carts have a rather simpler purpose, but toothed wheels arranged in chains of gears have been used as the mechanism of clocks for many centuries. Originally the motive power for clocks was derived from slowly descending weights, or sometimes from flowing water, but more recent clocks have used spiral springs wound up under tension or even electric motors. The whole point of a clock, of course, is that it should keep accurate time, and the pendulum has traditionally served to regulate the rate of motion of the whole gear chain driving the clock hands. As Galileo discovered, the swing rate of a pendulum is essentially independent of whether it makes large or small swings, and depends only upon the length of the supporting rod or chain, so that it is excellent for this purpose. A pendulum one metre in length takes almost exactly two seconds to make a complete forward and backward oscillation, so that it ticks in seconds. Shorter pendulums beat more rapidly, in proportion to the reciprocal of the square root of their length, so that a pendulum to tick twice a second (period one second) must be one quarter of a metre long. Pendulum clocks are, however, not suitable for carrying around, so that portable clocks and watches us a sort of rotating pendulum called a balance wheel, whose restoring force is provided by a fine spiral "hair spring". More recently, of course, even this has been replaced by the extremely rapidly vibrating wafer of quartz crystal found in modern electronic watches.

Modern machines are beautiful examples of precision engineering, and a well-designed machine is a thing of beauty as well as being extremely useful.

Clock makers of last century recognised this and often made the polished brass wheels of their clocks visible through the sides of glass cases. The machines in a modern factory have a similar aesthetic appeal.

Engines

An engine or a motor is a machine that takes in energy in some bulk form and produces ordered mechanical motion, generally as a rotating shaft, with enough power to drive other machines. Modern industrial society would be impossible without engines and, while we may sometimes regret the passing of farm horses in favour of tractors, no one could now seriously advocate the use of animals to provide mechanical energy in any more regimented fashion to drive machines, as once was done.

We can divide engines fairly generally into heat engines, in which some sort of fuel is burnt to provide a high temperature to drive the engine, and those such as electric motors and water wheels that work in some other fashion. Of course we might note that the electric power used in an electric motor may itself have been derived from a coal-burning thermal power station, and that the heat of the sun was ultimately responsible for driving the weather and causing rivers to run, but for the present we take a more local view. We saw above that heat engines are not very efficient—usually in the range from 20 percent for small engines to 50 percent for large ones—but electric motors may be more than 95 percent efficient. Water wheels and windmills are not usually very efficient, but special turbines in hydroelectric power stations can extract a large fraction of the energy from the water stream.

When the power output of an engine is in the form of a rotating shaft, it is simple to connect it so as to drive other rotating machinery. Some engines, however, work in quite a different way. An aircraft with propellers is very different from a rocket, and an aircraft with jet engines comes somewhere in between. Because rockets used to launch satellites are now commonplace, it is worthwhile to give brief attention to how they work. Basically what happens is that fuel is burnt at high temperature inside the rocket motor and the hot gas comes out from the tail of the rocket at high speed. By one of the "laws" worked out by Newton, the force acting on the hot gases to eject them is exactly balanced by an equal force acting on the combustion chamber from which they emerge, and this force is enough to push the rocket into the sky. There is no need for any surrounding air for the hot gases to push against, and indeed the rocket works better in empty space where there is no air resistance to overcome.

A jet engine on an aircraft works in something the same way, except that it has to take in air to burn the kerosene fuel, and it needs a turbine compressor rather like a small propeller inside the front of the engine to push the air into the combustion chamber. This small propeller provides part of the thrust of the engine, and there is a balance between this propeller or fan thrust and that from the pure jet part of the engine that varies from one design to another. Large civil airlines use fan-jet engines with quite large fans for efficiency, while military fighter aircraft tend to use nearly pure-jet engines for power. Both types can only operate within the atmosphere.

New types of engines and motors for all sorts of purposes are continually being developed. Between them they power our civilisation and allow us to lead the kind of life to which we have become accustomed.

Controlling Machines

Once machines had built-in engines, it became important to develop ways of controlling them—a runaway steam train can be a very dangerous thing! In most cases we get around this potential problem by having a human minder for the machine, and this is often the easiest thing. Sewing machines, kitchen mixers and motor cars have well defined tasks that they can do, but these vary widely in detail and a controlling intelligence is necessary. Electric generators or water pumps, on the other hand, can do only a single thing and require a minder only to ensure that nothing goes wrong.

The second and simpler sort of situation can be handled efficiently in two ways. First we can arrange for all the performance parameters of the machine—temperatures, pressures and so on—to be displayed somewhere convenient so that a human operator can make adjustments as necessary. Even better, we can arrange that, when one of the parameters strays too far from normal, the machine automatically makes an adjustment to itself to correct the situation. This sort of feed-back control mechanism was introduced on steam engines in factories some 200 years ago, and similar "governors" or controllers are now common in nearly all machines. Even things such as motor vehicles, which have a human driver, have all sorts of automatic feedback mechanisms to keep the fuel flow, engine temperature, battery voltage, and a host of other things close to their design values.

For the more general situation in which a machine might be called upon to do a variety of things, we need a much more sophisticated control arrangement which can be told in some general way what is required and which will then make all the necessary internal adjustments. Washing machines and microwave ovens now have microprocessors, which are small computer chips, to do just this, and it has become a commonplace of life. Much more complex control systems can carry out extremely difficult tasks such as navigating a large aircraft to its destination and landing it without human intervention, although this is not usually done in practice. Industrial robots can similarly assemble cars and test them on automatic production lines, taking much of the drudgery out of industrial life.

There is, of course, a social problem associated with these developments. In the days when only humans could make decisions—even trivial ones such as which bolt to tighten in an assembly line—there was a continuing demand for human minds for these purposes, even if the muscle power was provided by machines. Now that machines are sufficiently "intelligent" to make these decisions for themselves, humans are no longer required. But that leaves the problem of employment—how can the machine minders of past years continue to make a useful contribution to society? The answer is not one of the questions that science is called upon to answer, but rather one for which the social sciences have been developed. The answer clearly requires insight into the nature of society and the rights and responsibilities of individuals within it. It is not a question that will solve itself, nor is it one that can be safely ignored as the capabilities of intelligent machines increase. A Luddite solution—destroy the machines—cannot however be sustained in the long run, any more than can a return to human ecological balance through high infant mortality.

17

Communicating Information

"Where is the wisdom we have lost in knowledge? Where is the knowledge we have lost in information?" T.S. Eliot "The Rock"

The second half of the twentieth century is certainly the age of information and communication. While printed books and magazines continue to be produced in increasing quantities and serve to record most of our worthwhile information, radio and television penetrate every home and convey up-to-the-minute information (whether accurate or not!) on very nearly all subjects. Happenings deemed to be important, or rather "newsworthy" in a marketing sense, receive immediate coverage, and there is also a rather small fraction of information of more lasting value.

At a more pedestrian but perhaps more important level, information on bank transactions is flashed from automatic teller machines to central computers, stock exchanges record share transactions from around the world, and statesmen and armies keep continual watch on each other. Information has real value in time, in dollars and in lives.

There are two interlocking parts to this technology that affects all our lives. The first is the "hardware" or computer chips, telephone lines and communication satellites that makes it all technically possible; the second is the "software" which carries all the instructions to make these devices carry out the particular functions desired, out of all the host of other possible actions. On top of that, at least in the case of radio and television, there is a third "humanities" aspect that determines what is broadcast. It is perhaps sad that so much technical perfection is deployed for content as trivial as the average television program, but that is another story. Here we discuss in outline just some of the principal technical aspects of information and communication.

Information

It is interesting to see just what is meant by "information" in a technical sense, for it is rather different from what we might imagine. Fairly clearly information is rather different from the words used to convey it, so we want to try to define it more carefully. Two people exchange information by sending a message from one to the other. The number of possible messages they might exchange may be very large but is not infinite. We can then imagine all of the possible messages to be written in a huge book so that the information is actually sent by simply sending the reference number for the message.

Any particular message in the book can be located by sending a sequence of "Yes/No" bits of information in answer to the following questions which, by agreement, do not really need to be asked: "Is the message in the first half of the book?"; "Is it in the first half of that part?" and so on. By counting the number of questions necessary to find the message that is being sent, or equivalently the number of Yes/No bits that need to be sent, we have a measure of the information content of the message. Conventionally "Yes" is signified by sending the digit 1 and "No" by sending a 0, so that the message is a whole string of digits like 11011000101 and the information content in "bits" is just the length of the string. The word "bit", incidentally, is both a sensible term and also short for "binary digit".

This representation of information is important in communication problems and also for storage of information. A computer can easily store numbers in this form, because we can actually represent any number by a string of ones and zeros. We can do the same thing for letters of the alphabet, and thus for words, by noting that 128 characters is more than we have on a typewriter keyboard, counting both upper and lower-case letters, and we need only seven binary digits 1 or 0 to specify all numbers from 0 to 128. A compact disc stores music in just this form too, as tiny pits on the disc surface that are then read by a sort of laser microscope. To recreate the music, the laser must read about 700,000 bits of information each second. A television picture of ideal resolution requires nearly 100 million bits per second, but that would allow for a complete change of the picture every 1/25 of a second, which does not happen in reality, and the picture can be compressed into perhaps one tenth of that information.

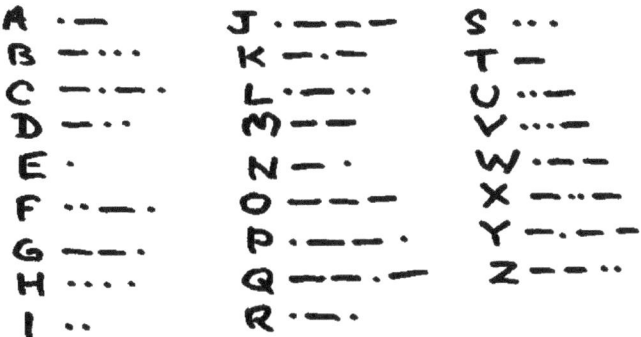

The Morse code used a series of short and long tones, called dots and dashes, to code all the letters and numbers. These could easily be sent along an electric circuit using a key switch.

Communications

Communications systems are of two basic types—broadcast systems such as radio and TV, and individual systems such as the telephone system. The technical problems are rather different in each case.

The first telegraph communication systems, or data networks as we would now call them, were chains of towers each equipped with two large hinged arms that could be moved to various positions to indicate the letters of the message being sent. An observer in each tower watched the arms of the previous tower through a telescope and recorded the message, which he then passed on to his own operator to retransmit to the next tower in the chain. The first of these communication lines was built in Europe about 200 years ago, and by 1850 there were lines across most of Europe. The French network alone had 556 stations and 4800 km of lines linking 29 cities to Paris. The need for retransmission at each station obviously restricted the speed of the network, but quite a lot of data could be passed by using an efficient code book.

The first electric telegraph was erected in America in 1844 by the American inventor Samuel Morse (1791–1872). It used copper wire and a key switch to spell out messages using a simple code of dots and dashes called the Morse code. Because it was simple and reliable, this system remained in use around the world for sending telegrams and other brief messages until about 1960, although human operators were increasingly replaced by automatic printers. Messages were sent between telegraph offices in each town and then delivered by hand by messengers—typically boys on bicycles in Australia.

The telephone was invented by the Scottish born American scientist Alexander Graham Bell (1847–1922) in 1876 and very rapidly came into commercial use because it was so much simpler and more direct than the telegraph. It probably did more to establish communication networks around the world than any other invention, particularly with the introduction of submarine cables linking the continents and copper wire cables to nearly every house in developed countries. The present telephone system is still similar in principle to early systems, but switch girls have been replaced by sophisticated electronic exchanges, long-distance cables have been replaced by microwave relays and then by optical fibre links, and submarine cables by satellites and then by optical fibres.

The existence of radio waves was predicted theoretically by the British physicist James Clerk Maxwell in 1873, they were produced experimentally by the German physicist Heinrich Hertz in 1888, and shown to be a practical means of communication across the Atlantic Ocean by the Italian engineer Guglielmo Marconi in 1901. Today radio waves of all types are used for sound and television broadcasts, local communication, and to study the stars and galaxies. The development of Earth-orbiting satellites in the 1960s, and particularly of geostationary satellites which make one orbit of the Earth in 24 hours and thus stay over the same point on the Earth's surface, made it possible to relay radio signals around the world without having to rely upon reflection from the variable and unreliable ionosphere. Satellite transmitters are also now used for direct rebroadcast of radio and television signals to domestic receivers equipped with dish antennas. A major problem with radio waves, however, is the sheer volume of information traffic that they carry. Because they are broadcast rather than confined to one line in space, interference is a problem and there has had to be international agreement about allocation of frequencies for competing services.

Television was invented by the British engineer John Logie Baird in 1926, but much of the development of the present system took place in America. Although television signals are broadcast, they require considerable power to give a good picture and therefore do not cover large distances in the same way that radio signals do. Television signals use rather short wavelengths, about twice the length of the bars on a television antenna, which do not propagate well into valleys and which are shaded by hills, so that repeater stations are necessary. Special microwave links are used to carry programs to the major transmitters in a network, although some of these have now been replaced by satellite links.

Communicating Information

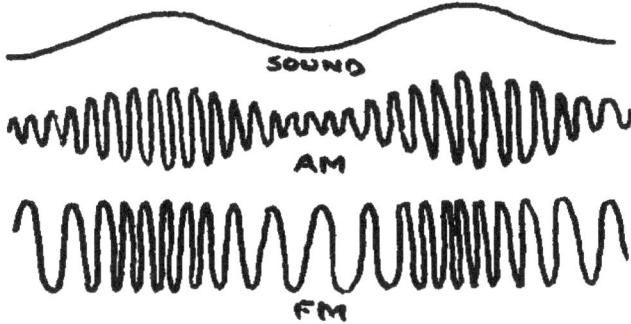

In an amplitude-modulated (AM) radio signal the energy of the wave is made to vary in the same way as the pressure in the sound wave being transmitted—the music or the announcer's voice. In a frequency-modulated (FM) radio signal the amplitude stays constant but the frequency is changed to follow the sound pressure signal.

On top of these communication links we now have a rapidly extending network of computer data links between government departments, the offices of major companies, and even private citizens. Much of the data exchanged may be unnecessary or trivial, as is true of telephone conversations, but the communication links across the world are gradually making us into what Marshall McLuhan called a global village. Perhaps ultimately we will manage to develop a global consciousness of our common problems and be able to work together to solve them.

Computers

The first person to design a practical computer was the English mathematician Charles Babbage (1792–1871) and the first computer programmer was his mathematical collaborator, Lady Ada Lovelace. As you can imagine, building a new and very complex machine in those days was extremely difficult, even though it consisted entirely of wheels and gears, and the final design was not completed but languished as drawings for more than a hundred years. A recent version built to his exact designs shows, however, that it would have worked just as intended.

Large computers, at first partly mechanical like telephone exchange equipment, were first built in the 1940s in England and America to help with codebreaking and other wartime activities. One of the best-known names is that of the Englishman Alan Turing (1912–1954), though many others contributed. The first civilian computers appeared in universities around 1950. These were huge machines, occupying several large rooms, and had a computing power about equal to that of a programmable pocket calculator of today. Nevertheless they performed immensely useful work and began the computer "revolution" in modern technology. The trend to miniaturisation of components began in 1950 with the invention of the transistor at the Bell Telephone Laboratories in the US but did not affect computers until the early 1970s. The crucial step was the development of integrated circuits, in which a whole computer processor, or a whole slab of memory, consisting at first of thousands and now of more than a million transistors with connecting wiring and other components, is printed photographically onto a wafer of silicon and processed through perhaps as many as a hundred steps to produce the final "chip". Not only did this reduce the size of computer components but, even more importantly, it reduced their cost by a factor now reaching about a million compared with the 1950s. The other thing that has changed is the basic speed of computing operations, initially thousands per second and now more than a thousand million per second!

Along with this development in hardware came a parallel development in software—the programs, or sets of instructions telling the computer how to perform specific tasks such as calculations or wordprocessing or picture drawing or data transfer. Software and programming languages are as important a part of the total computer development as is the hardware itself. Basically all that a computer can do is to read numbers from an input device such as a keyboard, perform the ordinary operations of arithmetic upon them, and copy the result to an output device such as a printer or a screen. All the wonderful things that we see on computer screens are simply the results of this process, and it is the program that determines how it should be done.

One of the interesting and important recent developments is that of "expert systems" in computers. These systems are designed to give advice on matters such as making decisions on farm management of forest fire control in which there are no exactly correct answers, only best strategies. The program incorporates a very large amount of empirical wisdom gathered by interviewing human experts in the field and recording what they advise and what factors they would take into account in reaching a decision. This expert data can be recorded in a form such as "If events A and B are observed, then the likelihood that C is the best thing to do is x percent." With all this information available, a computer can rapidly compare the data it has on a particular situation with the recorded advice of one or more experts and make a recommendation for action based upon the highest likelihood for the given situation.

Ordinary computers that follow programs in any of these ways are called "von Neumann machines" after the American mathematician John von Neumann (1903–1957) who worked extensively on computer programming. Such computers are completely logical and do exactly what they are told in sequential fashion. Even the "supercomputers" of today do just the same thing, except that they are extremely fast and may do many calculations at the one time. These computers can do wonderful things but their results are, in principle, exactly predictable.

Recently a new sort of computer has been developed based upon a quite different design concept called a neural net, which seeks to mimic the way in which neurons in the animal brain interact. As we noted in Chapter 13, these neurons are nerve cells which have many input connections from other cells and a single output which either fires a pulse or remains quiet depending upon the pulses it receives from the other cells to which it is connected. The whole computer, consisting of a large number of artificial neurons of this type, "learns", or rather is "trained", by reading a long sequence of inputs and giving outputs which are noted as being either correct or wrong. The computer then varies the notice it takes of particular internal connections on the basis of right or wrong answers. The whole thing is so complicated and "internal" to the computer that it is impossible to say exactly how it works. The results are therefore not exactly predictable and will sometimes be wrong, just as in the case of a human, but the computer can, at least in principle, learn from its mistakes and become steadily more capable. So far the new types of computer are in their infancy and are restricted to carrying out fairly simple tasks such as recognising words from a limited vocabulary as spoken by different people, but they open up very interesting possibilities.

There is an interesting problem about "artificial intelligence" of the kind that one or other of these computer types might be said to possess. The nub of the problem is to try to define what we mean by "intelligence" in a human sense and then to design a fair test for computer intelligence. The most satisfactory test that has so far been suggested is as follows. Shut a human in a room so that the only communication with the outside world is by means of a typewriter that can both send and receive messages. Shut the computer in a similar room with a similar typewriter connected to its input and output. You then proceed to ask questions or have conversations with the occupants of the two rooms my means of their two typewriters. The questions can be about anything at all—

mathematics, music, life or love. The computer is assumed to have been "told" about the test, so that it does not give the game away by always giving correct answers to mathematical problems! If, after as long as you like, you have been unable to determine which room contains the human and which the computer, then it seems reasonable to say that the computer has an "intelligence" that is equivalent to that of a human.

Appendix A: Scientific method

The aim of this Appendix is to provide some more detailed information that does not fit easily into the original plan of the book. It can be read independently of the main text or as a supplement to it.

The Scientific Method

The process of advancing scientific knowledge is often formalised by those who study the history and philosophy of science as the "scientific method". Formally, what is done is that a "hypothesis" is formulated—a clear statement which could be of the form "A always leads to B". Many careful experiments are then carried out in which A is made to happen and the experimenter looks to see whether or not B follows. If a single case is observed where B does not follow A, then the hypothesis is disproved. On the other hand, no huge number of observations in which B follows A can ever prove the hypothesis correct beyond doubt. Still, a hypothesis that has not yet been found to be in error has a lot going for it!

Even though a hypothesis has been formally disproved in its original form, an experienced scientist might still believe it to contain a useful idea. Perhaps a modified version such as "A leads to B except when condition C applies" would prove to be reliable. So then this new and more specific hypothesis can be tested.

An example of this is Newton's theory of mechanics, which works excellently provided the speeds of all the bodies involved are much less than the speed of light. If this is not the case, then the theory gives wrong answers, so that the hypothesis that it is a correct account of what is going on has been disproved. Einstein's theory of relativity, which is much more complex than Newton's theory of mechanics, fixes this problem. But for most ordinary events we still use Newton's theory, remembering to add the proviso "except when velocities are not small compared with the speed of light."

Of course, science would be a dull subject if everyone followed the "scientific method" all the time. In reality, the hypothesis being investigated is usually "Something interesting will happen if I do X" and this is nearly always supported by experiment! Insight then into the formulation and testing of hypotheses is a matter of imagination, guesswork and experience, and the only hypotheses that are worth the labour of experimental testing are those that seem likely to really tell us something interesting. But when it comes to the crunch, the formal scientific approach must be used to establish reliable knowledge, and often the knowledge gained by disproving a hypothesis is just as valuable as if all the experiments had supported it, for it tells us to look in another direction in order to understand what is going on.

In another picture, science is rather like a treasure hunt, and the scientist uses experience, guesswork and imagination to make a hypothesis like "I bet that the answer lies under that stone." Then we conduct the experiment of lifting the stone and looking under it. Sometimes we are lucky and get the treasure (knowledge and understanding, in this case) under the first stone, but sometimes it takes many attempts. Experience tells us which sorts of stones are worth looking under, so our skill grows with our experience. We might be a little disappointed if there is nothing under our stone, but that is all part of the game and we do not need to look under that particular one again. All this adds to the joy when we do at last find something!

Hypotheses and postulates

It is important to realise the distinction between a hypothesis such as this and what is called a postulate, which is really an unquestioned assumption. A good example of a postulate is Euclid's geometrical statement "Given a straight line, and a point not lying on that line, there is only one line through that point that does not meet the original line, no matter how much each is lengthened." This is the basis of ordinary "Euclidean" geometry. But we could have made the postulate that no lines can be drawn through the point without meeting the original line, and this leads to the sort of geometry that refers to patterns drawn on the surface of a closed object such as the earth. Alternatively, we could postulate that an infinite number of lines could be drawn through the point, and this leads to another sort of geometry called hyperbolic geometry, which is something that could be drawn on the doubly curved surface of a saddle.

To make this all clear, we generally need to define closely what we mean by various terms. In the present case, we must state that a "straight line" is the shortest curve that can be drawn between two points on the surface concerned, for example by stretching a piece of string between them. We then see that we need to define what is meant by a surface, namely a domain that contains just two dimensions, such as length and breadth, or longitude and latitude. A plane surface is just one particular example.

Definitions and laws

As noted just above, if we want to make any definite statement about anything, then we have to be quite clear what we mean by the words we use. This problem has bothered philosophers for three thousand years—if we ask "What is the meaning of Life?", a typical philosophical question, then we find that we don't really know what we mean by "Life" and the meaning of "meaning" is almost completely obscure! Even the word "is" presents problems. So we conclude that the question doesn't really mean anything at all, and even that is an advance in a philosophical sense!

Fortunately things are not so bad in science, since we are mostly concerned with things that have (or appear to have) a concrete existence in the observable world. We can therefore define time to be the thing that is measured by a clock that has been adjusted so that it marks just 24 hours in a day. Distance used to be defined in terms of marks engraved on a metal bar (the standard metre) preserved in a laboratory near Paris, and mass in terms of a standard platinum kilogram kept in the same laboratory. Standards laboratories around the world then regularly compared their own national standards of length and mass with those kept in Paris, so that components made to precise dimensions in one country would match exactly with those made in another. Now most of these standards have been redefined in terms of the properties of particular atoms, and therefore know no national boundaries, but National Measurement Laboratories in most countries have frequent intercomparisons of their measuring instruments for more complex quantities such as electric voltage or sound pressure.

Once we have these fundamental units of mass, length and time straight, we can then go on to define exactly what we mean by velocity, by force, by electric charge, and so on. When these definitions have been agreed and written down, it is possible to formulate precise statements of physical "laws", by which we mean hypotheses that have been tested often enough that we are pretty sure that they are reliable statements about the way the world behaves. In the physical sciences, most of these laws can be expressed as mathematical relations between some of the quantities that have been defined, for example "force equals mass times velocity."

Not all scientific understanding has the form of such quantitative "laws", however. Particularly in biology, many of the statements of reliable knowledge are in more qualitative form, as for example in clarifying the role of DNA in cell division and heredity, or the role of particular viruses in causing diseases. But once again, we must be quite sure what is meant by the words "DNA" and "cell" and "virus" before any such statements make clear sense.

Uncertainties and statistics

While it would be nice if every question we can ask in science had a clear and logical answer, many answers are not like that. Part of this uncertainty is inherent in nature—we may know that the half-life of a radioactive atom (the time after which half of these atoms will have decayed) is, say, 15 minutes, but, if we concentrate on the fate of a particular atom, it may decay after just 10 seconds or it may still be undecayed after an hour. According to our present understanding of radioactive decay, which is based upon quantum mechanics and appears to be completely reliable, there is not, and never can be, any way of predicting just when any particular atom will decay.

A normal-scale illustration of this principle is given by the tossing of a coin. The probability of it showing heads is 1 in 2 or 0.5, since both heads and tails are equally likely. (In this case, however, if we knew in detail exactly how the coin was thrown, we could calculate the result, but let us assume that this is not known, and that the person tossing the coin is not expert enough to influence the outcome.) Suppose we have tossed the coin five times and it has come down tails each time, what is the probability of heads on the next throw? The answer is that it is still 0.5, since each throw is independent of all others. Actually the probability of throwing five tails in a row in just five tosses is $(1/2)^5$ or 1/32 or about 3%, so it is likely to happen about once in every 30 attempts. The uncertainty in this estimate of occurrence is, however, quite large, and it is almost equally likely to happen either twice or not at all in 30 attempts. (The uncertainty is actually about equal to the square root of the number of expect successes, so the answer is 1 ± 1.) If we make 300 attempts, however, then we can expect 10 successes with an uncertainty of only about ± 3.

There is another sort of uncertainty that is not fundamental, like quantum uncertainty, or even probabilistic, like tossing a coin, but that is due to the difficulty of making precise measurements in many cases. If we measure a physical quantity X many times, using the best possible techniques, then the answers

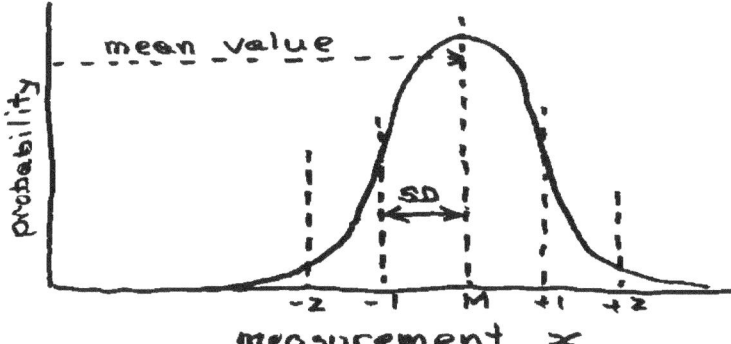

The normal distribution curve expected in many types of measurement. The height of the curve shows the relative likelihood of measuring a particular value. The mean is the average measured value, and the standard deviation (SD) is the uncertainty usually quoted. 95% of measurements lie within two standard deviations of the mean.

will vary just a little around their mean value. Usually this variation will be symmetrical about the mean value—just as likely to be a little above the mean as the same amount below—with the probability of obtaining a particular result following a bell-shaped curve known as a "normal distribution". This curve is shown in the figure. When enough measurements have been made, the value of X can be specified, for example as 12.3 ± 0.2 with the uncertainty being the "standard deviation" of the measurements. In a normal distribution, 68% of the measurements will lie within one standard deviation of the mean value, and 95% within two standard deviations of the mean.

Finally, there is the sort of uncertainty that is inherent in the thing being measured. Suppose, for example, that we ask the weight of a sheep or the number of red corpuscles per millilitre in human blood, then there simply is no exact answer. Instead we must usually be satisfied to know the average value of the quantity with which we are concerned and the extent of variation to be expected in that quantity across the population. This distribution may be quite different from a normal distribution, however, and might, in these two cases for example, show two peaks—one for males and one for females. One possible answer would be simply to give the mean value and the standard deviation, but a more careful presentation of the results should give the actual distribution curve.

An important use of statistical theory is in determining whether or not the result derived from some experiment really does support a particular hypothesis in a significant way. Returning to the coin-tossing example, suppose that we make the hypothesis that the coin actually has tails on both sides. We carry out an experiment by tossing the coin five times and each time it shows tails. Have we "proved" our hypothesis correct? Certainly not, since we know that such proof is impossible—there may be some other explanation. But to what extent does the result support the hypothesis? To examine this, we ask "What is the probability that this result could have occurred by chance?"—a question often dignified with the title "null hypothesis". In the case of the coin, we know that this possibility is only 3%, so that the experiment gives rather strong support to the original hypothesis. We would say that the experiment "supports the hypothesis at the 3% level" and generally support at better than the 5% level is regarded as fairly persuasive. Of course, the obvious thing is to do a further experiment. If we again get five tails in five tosses, then the likelihood of this occurring by chance, along with the same result in the original trial, is $(1/32)^2$ or $1/2^{10}$ since it is equivalent either to doing the 5-tosses experiment twice, or the single-toss experiment 10 times. The answer is 0.001 or 0.1%, which is so unlikely that we are justified in concluding "beyond reasonable doubt" (as they say in law) that the coin either has tails on both sides or is unbalanced in some other equivalent way.

Revolutionary ideas

Science progresses in two different ways that are quite different but complementary. The first, and most exciting, is the new insight that revolutionises the way in which a particular branch of science is viewed. Sometimes this may start with a new experimental observation, such as the discovery of radioactivity, and sometimes it may be a theoretical insight that lets a whole set of well-known observations suddenly make sense, such as Copernicus's realisation that the Earth and the other planets rotate about the sun. It might be in physics, like Newton's theory of gravitation, or the Schrödinger/Heisenberg theory of quantum mechanics, or Einstein's theory of relativity. It might be in biology, like Mendel's discovery of the laws of heredity, or Darwin's theory of evolution, or the discovery of DNA by Crick and Watson. It might be in geology, like the discovery of continental drift, or even in mathematics, like the development of chaotic dynamics. Sometimes these new discoveries mean that a whole branch of scientific understanding needs to be revised, but more often they start off an entirely new branch of science—the old theories still work well enough in their own domains, but now there is a whole new world opened up.

If this part of scientific progress is like opening a door into new room—what the historians of science call a "paradigm shift"—then the complementary part that follows is the careful exploration of the contents of that room. There are so many new consequences that follow from a new insight that it may take decades, or even centuries, to follow them through, though the pace of this exploration is quickening as it is gradually realised how important this new knowledge can be for all of humanity. Although very many scientists add considerable to our understanding of the world, only a few are fortunate enough to make a truly great discovery, and most of the scientific community is happy to spend their lives in the exploration process or in converting some of these discoveries to practical use.

Despite these exciting jumps in viewpoint, scientists are often thought by the community to be conservative individuals because they do not relish "new-age" ideas such as telepathy or crystal power or astrology. It is true that claims of these particular fields are indeed held in low esteem by scientists, but this is because they are based upon no experimental evidence and they do not represent any coherent and testable theory of what is going on. More than this, they are often promoted by using pseudo-scientific terminology to give a false impression of respectability. Perhaps the most interesting evidence is for UFOs ("unidentified flying objects"), but just because an individual is unable to identify something does not mean that it has any particular significance. I saw a UFO today, and it was actually a King Parrot, but my knowledge of bird species did not allow me to identify it!

Scientists would be delighted, however, if, for example, convincing and repeatable evidence could be produced for the reality of telepathic communication, for this would open an exciting new window on the human mind, and perhaps on physical processes in the universe, that could then be subject to detailed study. The same goes for predicting the future, or levitation, or the existence of ghosts. Unfortunately no such evidence has yet been produced for any of these claimed phenomena, and it is hardly worthwhile to expend effort on them while their claims are so dubious.

The community of science

Science is a cooperative venture, even though some scientists work alone, because we must share our findings and be able to rely upon things that are told us by other scientists. This second requirement imposes a rigorous morality on everyone who is involved with science—no telling lies! Anyone who transgressed this rule would be shunned by the entire scientific community, so it almost never happens. It is, of course, possible to be mistaken, and this is forgiven, though more than a single mistake would lead questions to be asked about one's scientific ability.

The other important thing is sharing, and the way in which this is done is through papers in scientific journals. Most researchers write just one or two papers per year, but when this is added up all round the world it comes to a great amount. The American Institute of Physics, for example, publishes nearly 100 different journals that together add up to more than ten metres of shelf space each year, and other branches of science are just as prolific. Chemistry is similarly unified, but biology tends to be divided into smaller societies concerned with particular branches of the subject. Before a paper is accepted for publication, the editor of the journal sends a copy to two or three world experts in the field ("referees") who make sure that it really does report something interesting, that there are no obvious mistakes, and that it is clearly written. Not all papers pass these tests, but those that do are then available to other scientists throughout the world. This publication process has been going on for more than 200 years and is the very backbone of science. More recently, of course, electronic means of communication on the Internet have become popular, but these essentially just make it easier and quicker to obtain access to papers that have otherwise been treated in the standard manner.

Ultimately the new knowledge that has been discovered and written down in all these hosts of papers becomes part of the generally accepted background of science. This knowledge is then distilled and crystallised to a form where it can be presented in a clear and coherent form in a book. If it is important enough, this book may ultimately become a text-book.

It is, of course, not just scientists who are interested in the findings of science—these are important for the general community as well. There are therefore many magazines, such as those listed in the Bibliography, whose mission is to present new findings in science in a form that can be fairly readily understood by ordinary educated people.

Appendix B: Scientific units

Over the ages, people investigating nature have used a vast array of different units to measure it. The British system of units is probably familiar to older people (or those who have lived in the United States, where it is still in common use). Lengths in this system are measured in miles, yards, feet and inches with a host of other units such as rods, furlongs and leagues for special purposes, while mass was measured in tons, hundredweight, pounds and ounces. The conversion between different units was complex (12 inches to the foot, 3 feet to the yard, 1760 yards to the mile, and so on). Only time was measured in what we now regards as standard units of seconds, hours, years. While this system is fine for everyday use, it is extremely complicated for scientific or engineering calculations, and there were hosts of secondary "derived" units such as acres for area, horsepower for power, and miles per hour for speed, the definitions of which all had to be memorised.

SI Units

In parallel with this, the metric system, based entirely on powers of ten and first suggested in 1585 by Simon Stevin, an inspector of dykes in the Low Countries, was formally introduced by the French in 1795. In a revised form, developed over the past fifty years, it is now in almost universal use in science with the title "Système Internationale" or simply "SI units". There are just a few fundamental units in the SI system: the kilogram (abbreviation kg) for mass, the metre (abbreviation m) for length, and the second for time suffice for mechanics. Compound units are defined to be various combinations of the basic units multiplied together or divided. Thus area, which is measured in square metres, has units written as m^2; volume, in cubic metres, is written as m^3; speed in metres per second as m/s or $m\,s^{-1}$; and so on. The unit of force is the newton (abbreviation N) which is the force necessary to accelerate a mass of 1 kg by 1 metre per second per second. (Interestingly, and as an aid to memory, 1 newton is about the weight of an average-sized apple!) The unit of work is the joule (J), which is the work done by a force of 1 N moving its point of application through 1 m, and the unit of power is the watt (W), which corresponds to performing 1 J of work each second. The unit of pressure is the pascal (Pa), which corresponds to 1 N m−2.

There are two things to notice immediately about these units. The first is that there are no awkward conversion factors: all the derived units come from simply multiplying or dividing the basic units. The second is a nice matter of terminology. Many of the units are named after famous scientists, Newton, Joule, Watt, and so on, and the abbreviation for the unit is a capital letter, N, J and W in these cases. But when we want to write the name of the unit in full, we use a small letter: newton, joule or watt.

When we deal with branches of science other than simple mechanics, we need some extra units. Thus simple electricity demands the volt (V), the ampere (A) and the ohm (Ω, the Greek letter capital omega). Things that vibrate need a unit to count the frequency, and this is the hertz (Hz), and so on. Note that these are all called after famous scientists of the past. Magnetic fields need new units, and so do quantities such as the magnitudes of capacitors and inductors that go into radio circuits. When we get even further from simplicity, we have units for radioactivity, for the huge distances to the stars, and so on. But it all fits into a nice unified system.

Common SI Units and Symbols

length	metre	m	force	newton	N
mass	kilogram	kg	pressure	pascal	Pa
time	second	s	electric charge	coulomb	C
frequency	hertz	Hz	electric potential	volt	V
work	joule	J	electric current	ampere	A
power	watt	W	magnetic field	tesla	T

The basic units are fine for many ordinary tasks, but often we need to measure something very much smaller or very much larger than the basic unit. In the SI system, the basic factor is 1000, rather than 10, because there is such a huge range of magnitudes to be spanned. The prefixes that indicate the multiplier are as shown in the table (the symbol μ for micro is the Greek letter mu). Just a few multipliers other than powers of 1000 have survived: those for 100, 10, 1/10 and 1/100. The complete range is shown in the table.

A few multiples of the basic units have been given special names because they are in such common use. Thus a common unit for area is the hectare

Appendix B: Scientific units

Multiples and Submultiples

name	deka	hecto	kilo	mega	giga	tera	peta	exa
symbol	da	h	k	M	G	T	P	E
factor	10	10^2	10^3	10^6	10^9	10^{12}	10^{15}	10^{18}
name	deci	centi	milli	micro	nano	pico	femto	atto
symbol	d	c	m	μ	n	p	f	a
factor	10^{-1}	10^{-2}	10^{-3}	10^{-6}	10^{-9}	10^{-12}	10^{-15}	10^{-18}

(1Ha=10,000 m^2), one for volume is the litre (1L=0.001 m^3 — the official symbol for the litre was originally a lower case l, but this can be confused with a one, 1, so upper-case L is often used), and one for mass is the tonne (1000 kg). All these multiples are, however, simple powers of 10.

The two misfits are time and angle, which retain peculiar multiple units (such as hours and weeks) or even basic units (degrees) that have been with us for a very long time. There are, however, good practical reasons not to tinker with these, and they do not really cause any confusion. (Angles in science are generally measured in radians, of which there are 2π in a full revolution.)

Special cases

There are a few special values that are worth remembering. The first is the density of pure water, which is just about $1000\,\text{kg}\,\text{m}^{-3}$ at 4°C, a figure that is the same as $1\,\text{kg}\,\text{L}^{-1}$ or 1 g cm-3. Most solid and liquid materials range in density from about half this value (for light wood) to as much as 14 times for mercury. Another useful value to have in mind is the density of air, which is just about $1\,\text{kg}\,\text{m}^{-3}$ at normal temperature and pressure. Hydrogen has only 1/7 of the density of air and helium about 1/4, which explains why these gases are used in balloons. The hot air in hot-air balloons has about 3/4 the density of the surrounding air, but the difference is still great enough to lift the balloon with its gondola and passengers if the balloon itself is big enough.

Another important quantity is the speed of sound, which is about $340\,\text{m}\,\text{s}^{-1}$, though it depends a little upon temperature. The ratio of the speed of something to the speed of sound in the place where it is moving is called the Mach number (after the German scientist Ernst Mach). Commercial jet aircraft fly at about 0.9 of the speed of sound, and thus at Mach 0.9, while the Anglo-French Concorde aircraft cruise at about Mach 2, and military aircraft sometimes go even faster. Knowing the speed of sound allows us to estimate how far away a lightning flash is. We just count the number of seconds between seeing the flash and hearing the thunder and then divide by 3 to get the distance in kilometres.

Another very special case is the speed of light in empty space, which is just about $3 \times 10^8 \,\mathrm{m\,s^{-1}}$ or about 1000 million kilometres per hour. This speed, which is given the symbol c, figures in Einstein's famous formula $E = mc^2$ which says that a mass m in kilograms, if converted entirely to energy, would give mc^2 or about $10^{17}m$ joules. This is a tremendous amount of energy, and even violent nuclear reactions convert less than one part in a thousand of the available mass into energy. The speed of light is a little less when it is passing through a material such as glass, the decrease being typically to about 70% of its vacuum value but depending on the refractive index (optical bending power) of the material.

Appendix C: Science and Ethics

Science and technology have so great an influence on our lives that it is inevitable that ethical questions should be raised about many scientific and technological activities. Unfortunately much of the debate is obscured by ignorance and prejudice, so that it is important to persuade people to stand back and examine the facts in a dispassionate manner before coming to any conclusions.

One of the prime problems is that science, and its offspring technology, place great power in the hands of those who use them for their own ends, be these good or bad. It is true, for example, that scientists discovered details of nuclear reactions and from these the way in which to make a nuclear explosive, but it was a politician with no scientific training (President Truman) who decided that the atomic bomb should be dropped on Japan. More modestly, science and technology discover new ways to improve foods or to combat diseases, but it is accountants and business executives who decide whether or not they should be manufactured and at what price they should be sold. Should scientists refrain from investigating new fields because of fear that what they discover could be used to harm rather than to help?

The view of most scientists could be summarised by the aphorism "Knowledge is better than Ignorance". The contrary view would be hard to sustain in any rational argument, and the modified contrary "Ignorance is Bliss" has at best a temporary validity. Those who ultimately make decisions on matters involving new science or new technology, therefore, need to understand at least its major implications, and the foundations of that desire for understanding are laid at primary school.

In this appendix we shall examine just a few major ethical questions that have arisen because of recent scientific advances. "Recent" here is an important word, because older science and technology, such as the aerodynamic studies that led to the development of aircraft, have become assimilated into our culture, and few people would argue that aircraft should be banned because they can be used for war as well as for peaceful purposes.

Nuclear Power

The supply of electrical energy is crucial to the continued life of humanity. The burning of fossil fuels such as coal and oil pollutes the atmosphere, and their supply will in any case be exhausted before too long—probably within the lives of the present generation of school children. Solar energy of one type or another would be a good solution in the longer term, but is nuclear power a sensible shorter term possibility?

The immense power of nuclear explosives and the fact that they can leave behind radioactive contamination has made many people very wary of the development of nuclear power stations. This concern has been magnified by the few accidents that have occurred in existing power stations, notably Three Mile Island in the USA and Chernobyl in the former USSR. In neither case was there a nuclear explosion—indeed no nuclear reactor has ever exploded, and such an explosion is impossible in most of them—rather, an engineering fault or operating error caused the reactor core to overheat and "melt down". Such an accident, like all severe fires, causes immense damage in its vicinity, but the real concern is the release of radioactive smoke and gases. No such release occurred at Three Mile Island, because the reactor was enclosed in an airtight steel and concrete building specially designed contain any such accident. This type of steel containment vessel, the size of a large multi-storey building, can be seen around the research reactor at Lucas Heights near Sydney.

The Chernobyl reactor, on the other hand, had been cheaply built and had no such containment structure, so that radioactive products were able to escape and contaminate a large area of Europe. Statistics are difficult to find, but one would expect the contamination to lead to an increase in the incidence of certain types of cancer in the population. One advantage of nuclear poisons, however, is that they do not last for ever, unlike many chemical poisons. The rate at which radioactive materials decay to essentially harmless products is roughly proportional to their radioactivity, so that after a few years radioactive levels have decayed to nearly their background values. It is important to remember that there is natural radioactivity everywhere, simply from minerals in the soil and from cosmic rays that hit the Earth from outer space. Going into a cave, or even into a building basement that does not have adequate ventilation, may increase your rate of exposure to radioactivity by a very large factor. Even lying on a sunny beach that has a significant concentration of rare earth elements

Appendix C: Science and Ethics   169

increases your exposure!

What has been learnt from these very few nuclear accidents is that power reactors must be designed and built with adequate containment vessels, despite the added cost (accountants note!), and must be operated under strict safety standards.

Another concern is terrorism, and the fear that enriched uranium could be stolen and made into a nuclear bomb. That is indeed possible, and security precautions are clearly necessary, as they are also necessary in the explosives stores of mining companies, but possessing the uranium and building a bomb are two very different things. Biological weapons would be a much simpler alternative for the would-be terrorist.

I do not want to suggest an answer here to the question of use of nuclear power. Australia is fortunate in its reserves of coal and sunshine and is free from the industrial pollution of the northern hemisphere, so for us the question is not urgent, but that may not be true for the world as a whole. Certainly we owe it to ourselves, and to the rest of humanity, to put greater effort and resources into the development of non-polluting renewable energy sources, and that means essentially the using of solar power through direct conversion to electricity, wind power, tidal power, biomass generation, or chemical dissociation.

Genetic Engineering

In the biological sciences too, knowledge is power, and power can be used for whatever ends we will. The study of viruses and bacteria for the control of diseases is clearly admirable, but the information and techniques could also be used to manufacture and distribute harmful disease organisms through water supplies. Here too there is the possibility of accident, and the accidental release of toxic materials from a factory can cause immense damage and destruction. Indeed, more people have been killed in the accidental release of chemicals from factories in the past twenty years than in all the nuclear accidents that have ever occurred.

Apart from this, there is the ethical question of manufacture and sale of harmful substances. Hard drugs such as heroin have clearly addictive and damaging effects, and governments are rightly tough on the criminals who exploit the addiction they cause. But what about the addictive drug nicotine, as found in cigarettes? This contributes to hundreds of thousands of painful deaths from lung cancer each year, but some large companies make immense profits from it manufacture and sale. Science no longer enters the picture, so it is financiers and politicians upon whom the ethical burden falls.

Of immediate interest is the question of genetically modified plants and animals for food, a topic that is currently exciting a great deal of discussion. Of course, humans have been genetically modifying plants and animals for tens of thousands of years by selective breeding and hybridising. Some of the products are sterile and cannot produce viable offspring—for example mules, which are the offspring of a horse and a donkey—or revert to earlier types if allowed to reproduce, as do some hybrid plants.

Scientists have recently learnt, however, how to transfer specific genetic material directly from one organism to another in order to produce a desired effect such as resistance to a specific disease, instead of relying upon the statistical outcomes of selective breeding. This allows a much greater degree of control of outcomes and has the potential to produce crops with greater yields, better disease of pest resistance, perhaps even better flavour! But there are potential problems. Can the growth and spread of these new plants be controlled, for instance, particularly if they have been engineered to be herbicide resistant so that weeds can be more easily controlled?

Some people have concerns about eating genetically modified foods, despite the fact that their ancestors have been doing it for tens of millenia. Of course any new type of plant needs to be carefully checked before eating it—many things found growing in nature are quite poisonous, and just because something looks like a fruit you know does not mean that it really is! The same goes for animals—a puffer fish is said to be delectable to eat, but if prepared by an unskilled chef, the poison may kill you! So it is with genetically modified foods, though if the nature of the genetic modification is known then a good prediction can be made.

Cloning

The process of cloning involves taking cells from an adult plant or animal and causing them to develop into another individual that will have exactly the same genetic makeup as its single parent. This contrasts with normal reproduction in which the offspring inherits genetic material from both its parents and the number of possible combinations of chromosomes, and thus of genetically different offspring, is very large.

Gardeners, of course, have been cloning plants for centuries. In the simplest process they damage the bark of a low branch and bend it down into the soil so that it grows roots. The branch can then be removed and grown as an individual plant identical to its parent. Another process is that of grafting, in which a small part of an adult plant is removed and surgically grafted onto the rootstock of a plant often of a different species. The new plant is then only a partial clone and has been genetically engineered, in a macroscopic sense, so as to have improved growth properties.

Only recently have similar things become possible with animals. In current techniques, the nucleus is removed from an egg cell and replaced with a nucleus taken from an adult cell of another individual of the same species. The egg is then reimplanted in its surrogate mother and allowed to develop into an individual which, when born, will be genetically identical with the animal from which the adult cell nucleus was taken. (One might speculate that it would have been easier to do this with a fish egg or a bird egg for which reimplantation would not have been necessary, but then there are other difficulties.)

Ethical problems apparently do not arise until the animal concerned is a human being, though some such worries should also be expressed in the case of other "higher" animals. It is not so much the actual cloning of a human being that presents the ethical problem, but rather the possibility that human embryos could be produced and then sacrificed to produce "spare parts" such as stem cells for use in surgical procedures. At what stage of development does an embryo acquire the status of a human being and thus the rights of protection under the law? This is not a scientific question, though science can perhaps provide some relevant information, but rather one to be decided by politicians after considering all the advice they can receive. This book is not the place to discuss these questions further; rather such discussions and decisions will take place in the general community. It is important that they are based upon accurate scientific information.

Bibliography

This annotated bibliography suggests sources for further reading that should be at an appropriate level for Teacher Education students. The number of books and journal articles that could be cited is immense, so that this represents a very small selection. Broadly speaking, I have suggested science texts that are written for upper high-school or beginning university students, supplemented by a few reference books and some books for the general reader and some well-known scientific magazines. The selection is quite limited, and there are many other fine books in libraries everywhere that will doubtless be suggested by lecturers.

With the recent importance of Internet-based information, we also give URLs for resources compiled by the Australian Academy of Science that have received international attention for their quality. Again, there are many other web sites that will doubtless be suggested by lecturers.

Australian school science texts

- *Primary Investigations* 7 volumes with 6 student workbooks. (Australian Academy of Science, Canberra 1994). A complete 'whole-school' K–6 approach to primary education in science, technology and environment. The course is experiment-based, using simple materials, and the main volumes are teacher resource books.

- *Perspectives of the Earth* (Australian Academy of Science, Canberra 1983). A descriptive text on all branches of Earth science, including astronomy, groundwater, and resource geology, as well as geology proper. Examples are drawn from an Australian context where possible. The book was written for senior high-school students.

- *Elements of Chemistry: Earth, Air, Fire and Water* (Australian Academy of Science, Canberra)

- *Biology: The Common Threads* 2 volumes (Australian Academy of Science, Canberra). A comprehensive text on all aspects of biological science. it includes environmental and ecosystem biology, cellular biology, disease and genetics. Examples are drawn from Australia where possible. The books were written for senior high-school students.

- *Biological Science: The Web of Life* (Australian Academy of Science, Canberra). Biological science emphasising an ecosystem approach, and with Australian examples. Written for senior high-school students.

Scientific magazines for the general reader

- *Nature*. A top weekly journal used for the rapid publication of news of important scientific advances. It also has articles of general interest.

- *Science*. Another prestigious weekly journal that contains both news of scientific advances and articles of a more general nature on scientific topics.

- *New Scientist*. A weekly scientific news magazine that describes recent advances and policy matters in science. Designed for the general reader interested in science.

- *Scientific American*. A monthly scientific magazine that presents longer articles for the general public on scientific subjects.

Books for the general reader

- *The* New Scientist *Inside Science* ed. Richard Fifield. (Penguin Books, London 1992). Twenty six important topics in physics, chemistry and biology, explained for the general reader.

- *The Mind of God* Paul Davies. (Penguin Books, London 1992). One physicist's view of the ultimate nature of things, and of the relation between scientific theory and reality. Paul Davies has written several other books on similar themes.

- *The Selfish Gene* Richard Dawkins. Plants and animals viewed as vehicles for preserving and multiplying genes. A perceptive and unusual view of genetics.

- *The Blind Watchmaker* Richard Dawkins. An excellent account of the modern biological view of evolution.

- *Chaos: Making a New Science* James Gleick. (Cardinal, London 1987). A popular exposition of the new science of complexity, fractals and chaotic dynamics.

- *Structures: Or Why Things Don't Fall Down* J. Gordon (Penguin, London 1978). An entertaining and informative look at buildings and other structures.

Web sites

With the increasing popularity of the Internet, and ready access to it, it is now possible to read and download material form a wide variety of sites. Of course, it is so simple to post material on a web site that much of what is available is rubbish, so that some sort of guide is required. The web browser google.com is one such guide, since it leads you to sites that have been rated for quality in at least a sensible though automatic way.

The Australian Academy of Science maintains a web site devoted to science education at primary and secondary levels. The URL to access this site is www.science.org.au/scied/index and this gives pointers to various resources. The following are of particular interest in the present context.

- *Good Science Books for Children* gives an extensive annotated bibliography in all branches of science, classified for the age groups 3–6, 5–8 and 7–12.

- *Back to Basics* gives an annotated list of URLs providing basic information on all branches of science at a level suitable for teachers as well as older children.

- *Primary Investigations*, with the URL www.science.org.au/pi, gives access to a range of material supporting the Academy's Primary Investigations course. Included are resources linking the presentation in *Primary Investigations* with State Education Department syllabus documents in New South Wales, Victoria, Queensland, and Western Australia.

- *NOVA: Science in the News*, which is designed for high-school teachers, advanced school pupils, and the general public, provides reliable information on a wide variety of science-related topics that have come to recent public attention. This site, with URL www.science.org.au/nova, also gives links to related URLs that have been examined and found to provide reliable and helpful information on the topic concerned.

- CSIRO has a great range of activities in science education. It maintains nine Science Education Centres around Australia, and publishes the science magazine *Scientiffic*, which has sections for children of all age groups from 7 years upwards and also has a Teachers' Guide. CSIRO is also the home of the Double Helix Club, which has some 11,000 young people as members. Information on these and other activities can be found at www.csiro.au/index.asp?type=educationIndex

- The Australian Broadcasting Corporation (ABC) has interesting and up-to-date programs on many branches of science. These programs and also news items on science are accessible through its web site:
www.abc.net.au/science/sitemap.htm

- The National Science and Technology Centre (Questacon) in Canberra has an excellent and changing exhibition of things relating to science, directed particularly at school students. Information is available on the site www.questacon.edu.au/index_flash.html

Index

aluminium, 90
amino acids, 94
analytical science, 4
animals, 111
    behaviour, 117
    brain, 116
    communication, 117
    evolution, 111
    food chain, 120
    habitat, 119
    nerves, 116
    size, 138
    systems, 113
    varieties, 113
Antarctica, 54
ants, 117
arch, 131
Aristarchus, 36
artificial intelligence, 153
atmosphere
    circulation, 57, 58
    pollution, 64
    pressure, 58
atoms, 15, 74, 75
    formation, 75
    structure, 80, 81
aurora, 69
Australian Academy of Science, 172, 174
Australian Broadcasting Corporation, 175

Babbage, Charles, 151
Baird, John Logie, 150

baking powder, 82
ball, bouncing, 17
beams, 133
Bell, Alexander Graham, 150
big-bang theory, 42, 75
binary numbers, 148
bits, 148
boiling, 15
books, 7
    for children, 174
    for the general reader, 173
    text, 172
brain, 116
bronze, 89

cancer, 7, 98
carbon dioxide, 63, 76, 105
cathedrals, 131
CD, 148
cell
    germ, 99
    simple, 96
ceramics, 88
chance, 158
chemicals
    pharmaceuticals, 85
    synthetic, 84
chemistry, 76
chlorophyll, 105
climate, 62
climate change, 62, 63
clock, 18, 143

Index

cloning, 170
    animals, 171
    humans, 171
    plants, 170
cloud droplets, 61
clouds, 60, 61
coal, 51, 109
colour, 29–31
communications, 149
complexity, 10
computers, 151
    artificial intelligence, 153
    expert systems, 152
    hardware, 151
    neural net, 153
    software, 152
concrete, reinforced, 135
consciousness, 117
continents, 45
cooking, 81
Copernicus, 37
Crick, Francis, 97
crystals, 77
    form, 79
    growth, 78, 80
CSIRO, 175
Curie, Marie, 5, 80
Curie, Pierre, 80
cyclones, 60

Dalton, John, 74
Darwin, Charles, 4, 100
definitions, 157
Democritus, 74
design, 91
detergent, 83
DNA, 97
double helix, 97
drugs, 169

Earth, 36
    age of, 44

    crust, 45
    formation of, 45
    magnetic field, 68
    magnetism, 67
    orbit, 37
    seasons, 39
    structure op, 45
    temperature, 56
earthquakes, 47
eclipse, 40
ecology, 10, 119
ecosystem
    managed, 123
    overpopulation, 123
ecosystems
    managed, 122
efficiency, 23, 141, 142
Einstein, Albert, 2
El Niño, 62
electric circuit, 70, 71
electric light, 71
electric motor, 72
electric power, 72
electric shock, 69
electricity, 66, 69
    alternating current, 72
    conductors, 70
    direct current, 72
    insulators, 70
    static, 69
electron, 70, 71, 80
element, 75
elements
    symbols, 75
emulsion, 83
energy, 13
    chemical, 18, 19
    conservation of, 23
    efficiency, 23, 142
    electrical, 19, 20, 73
    food, 19
    from plants, 109

gravitational, 16
heat, 14, 16
human requirements, 126
hydroelectric, 128
kinetic, 13
light, 21
nuclear, 20, 126, 168
other sources, 128
potential, 16
renewable, 169
solar, 20, 33, 104, 127
sound, 21, 26
stored, 16
thermonuclear, 127
transformation, 17
engineering, 6
genetic, 169
engines, 144
equipment, 9
erosion, 48
ethics, 6, 10, 167, 169–171
evolution, 100, 111, 112
time scale of, 102
experiments, 3, 8
impossible, 11
in biological science, 10
in physical science, 8
eye, 29, 30, 100

fabrics, 91
Faraday, Michael, 19, 71
feminism, 5
foams, 91
food, 125
genetically modified, 170
food chain, 120
food chemistry, 81
fossils, 11, 50, 101
frequency, 27
fuel
fossil, 51, 126
fundamental particles, 9

galaxy, 41
Galileo, 2, 38, 143
gas, natural, 51
genetic engineering, 108, 169
genetic variation, 100
genetically modified foods, 170
genetically modified plants, 170
geology, 44
glass, 79
gods, 14
Gondwana, 46, 47
greenhouse effect, 56, 63

hearing, 26, 27
heat, 14, 16
Hertz, Heinrich, 27, 150
holistic science, 4
humans, 102, 112, 115, 122
hypothesis, 156, 160

ice ages, 63
ice caps, 53
ideas, revolutionary, 160
ignorance, 167
information, 147, 148
infrared, 56
iron, 89
isobars, 60
isotope, 75, 80

jet engines, 144
joule, 16
Joule, James, 15
journals, scientific, 7, 162

Kepler, Johannes, 37
kinetic energy, 13
knowledge, 167
knowledge, reliable, 3, 6

lasers, 9
laws, scientific, 3
lens, 33, 34

Index

life, 43
    meaning of, 157
    on other planets, 103
    origin of, 11, 94
    what is it?, 93
light, 21, 22, 28, 32
    speed of, 2, 166
light year, 41
lightning, 70
liquids, 15
Lovelace, Ada, 151

machines, 140, 143
    controlling, 145
    effectiveness, 141
    efficiency, 141, 142
    replacing humans, 146
magazines, scientific, 173
Magellanic Clouds, 41
magnetic fields, 66, 68
magnetic poles, 67
magnetism, 66, 67
Marconi, Guglielmo, 150
materials
    composite, 134
    compressive strength, 130
    density, 130
    metals, 134
    tensile strength, 132
materials science, 87
Maxwell, James Clerk, 150
McLuhan, Marshall, 151
meiosis, 99
melting, 15
membrane, 96
metals, 51, 79, 89, 134
meteorology, 55
method, scientific, 3
metric system, 163
Milky Way, 41
mind, 117
minerals, 50

mirrors, 32
mitosis, 98, 99
models, 1, 2
molecules, 15, 19, 74, 76, 77
    biological, 95
momentum, 14
moon, 40
Morse code, 149
Morse, Samuel, 149
musical instruments, 26
mutation, 97

National Science and Technology Centre, 175
nerves, 116
neutron, 80
Newcomen, Thomas, 16
Newton, Isaac, 2, 14, 29, 38
normal distribution, 159
NOVA: Science in the news, 174
nuclear energy, 20
nuclear power, 168
nucleic acid, 96
nucleus, 80

oceans, 53
    circulation, 62
Ockham's razor, 3
oil, 51
ozone, 65

paradigm shift, 161
particles, fundamental, 9
patterns, 13
pendulum, 18
periodic table, 75
perpetual motion, 23
pharmaceuticals, 110
photosynthesis, 105
planets, 39
    motion of, 2
plant breeding, 107

plants, 104
  leaves, 106
  photosynthesis, 105
  reproduction, 107
  roots, 107
  trunk, 107
plastics, 90
politicians, 5, 6
polymers, 90
population, 120, 121
  human, 123
postulate, 156
power
  definition, 16
  human, 115
  mechanical, 144
  muscular, 115
  nuclear, 20, 168
  solar, 20, 169
predictions, 3
  testable, 2
Primary Investigations, 172, 174
proton, 80
Ptolemy, 36
publication, 3
pulsars, 42

quantum mechanics, 2
quark, 80
quasars, 42
Questacon, 175

radio, 24, 150
  AM and FM, 151
radioactive dating, 44
radioactivity, 75, 168
raindrops, 61
reductionism, 4
relativity, 2, 42
reliable knowledge, 3, 6
reproduction
  asexual, 98
  sexual, 99
rockets, 144
rocks, 44, 49
  igneous, 49
  sedimentary, 49
Roentgen, 9

satellites, 55
Savery, Thomas, 16
science
  analytical, 4
  biological, 10
  community of, 162
  ethics, 6
  holistic, 4
  laws, 157
  nature of, 1
  physical, 8
  practical, 5
  women in, 4
scientific method, 3, 155
sea level, 64
seasons, 39
self-organisation, 95
soap, 83
solar system, 38
solids, 15
sound, 21, 24–26
  speed of, 165
spectrum, 29
stability, of structures, 136
stars, 43, 75
statistics, 10, 158
steam engine, 19
steel, 89
strength
  compressive, 130
  tensile, 132
structures, 130
  beams, 133
  biological, 137
  stability of, 136

Index

stone, 131
stone arch, 131
strength, 130
sugar, 105
sun, 38, 56
    eclipse, 40
system
    digestive, 113
    nervous, 116
    respiratory, 114

technology, 5, 6
telephone, 150
telescope, 33
television, 150
theories, 2, 3
tides, 53
tone quality, 27
tools, 140
truth, 1, 8
Turing, Alan, 152

units
    multiples, 164
units of measurement, 163
universe, expanding, 42
universe, future of, 43
universe, origin of, 11

vesicle, 96
virus, 86
vision, 28, 30
voice, human, 28
volcanoes, 47
von Neumann, John, 153

water, 52, 76
    artesian, 52
    supplies, 125
Watson, James, 97
watt, 16
Watt, James, 16
waves, 21, 22
    electromagnetic, 22
weather
    fronts, 60
weather map, 59, 60
web sites, 174
Wilkins, Maurice, 97
women in science, 4

X-rays, 9

yeast, 82

www.ingramcontent.com/pod-product-compliance
Lightning Source LLC
Chambersburg PA
CBHW060946170426

43197CB00031B/2983